SpringerBriefs in Molecular Scie

Green Chemistry for Sustainability

Series Editor
Sanjay K. Sharma

For further volumes:
http://www.springer.com/series/10045

Bradley Ladewig · Benjamin Asquith

Desalination Concentrate Management

 Springer

Bradley Ladewig
Department of Chemical Engineering
Monash University
Clayton, VIC 3800
Australia
e-mail: bradley.ladewig@monash.edu

Benjamin Asquith
Department of Chemical Engineering
Monash University
Clayton, VIC 3800
Australia
e-mail: Ben.Asquith@monash.edu

ISSN 2191-5407
ISBN 978-3-642-24851-1
DOI 10.1007/978-3-642-24852-8
Springer Heidelberg Dordrecht London New York

e-ISSN 2191-5415
e-ISBN 978-3-642-24852-8

Library of Congress Control Number: 2011940848

Printed on acid-free paper

Springer is part of Springer Science+Business Media (www.springer.com)

Acknowledgments

The authors gratefully acknowledge the financial support of the National Centre of Excellence in Desalination Australia

Contents

Chapter 1
Introduction

Water demand is rising globally, largely driven by increases in population and living standards. This includes water used for domestic human purposes, including drinking, washing, bathing and household purposes, as well as industrial and agricultural use. However due to changes in the spatial distribution of the demand, for example the large and growing demand from cities in coastal regions around the world, the ability to meet the demand from conventional sources is being severely stretched. Further exacerbating the problem is the increasingly variable nature of rainfall events, that is extended periods of drought followed by severe flood events. This phenomenon is predicted to increase as a result of global climate change, and presents new and challenging aspects to the use of rivers, lakes and dams for water supply.

The response to these coupled challenges of increasing demand and decreasing availability of conventional supplies has generally been to implement water conservation strategies while at the same time seeking new sources of water. Perhaps the most obvious source of additional supply, particularly for coastal cities, is desalinated seawater. The world's oceans cover more than half of the world's surface and account for almost 97% of all water on the planet, however the main impediment to their use for human purposes is the high salt content. To render seawater or other saline water suitable for human, industrial or agricultural use it must be desalinated or desalted, that is, have the salt removed from it. This is actually a somewhat generalized term, as much more than just sodium chloride or salt needs to be removed from seawater to make it suitable for use. However the term is well accepted to mean the treatment of a water source containing salts and other contaminants, to produce a product fit for a specified purpose—often human consumption.

Desalination has been practiced for many years, although the early large-scale industrial applications were in the 1960s in the Middle-East using thermal desalination technologies such as multistage flash or multi-effect distillation. These plants used sophisticated thermal or vacuum systems to effectively distil a portion

B. Ladewig and B. Asquith, *Desalination Concentrate Management*,
SpringerBriefs in Green Chemistry for Sustainability,
DOI: 10.1007/978-3-642-24852-8_1, © The Author(s) 2012

of the feed, producing highly pure water. These plants had quite high energy consumption relative to the quantity of product water produced, however the abundant local supply of gas and oil (for producing the heat required to drive the processes) meant that these technologies were competitive, and could be deployed on very large scales. The by-product of these processes, which often operated with only around 30% recovery, is concentrated seawater or brine, hereafter referred to as concentrate.

More recently developments in polymer membrane materials and technologies have driven the adoption of pressure-driven desalination as the preferred and most cost-effective desalination technology. In fact almost all new large-scale seawater desalination plants now employ membrane technologies, such as micro- and nano-filtration for pre-treatment followed by one or more reverse osmosis stages. These plants usually operate with around 50% recovery, meaning that the by-product is a concentrate with approximately double the salt concentration of the feed water. To illustrate this shift from thermal to pressure-driven desalination processes, in 1999, reverse osmosis and multi-stage flash processes accounted for 10% and 78% of the world's desalination capacity respectively. However, in 2008 reverse osmosis processes accounted for more than half of the world's desalination capacity (Economic and Social Commission for Western Asia 2009). For a comprehensive description of the different desalination technologies, please see the reviews of Fane et al. (2011) and Khawaji et al. 2008).

In previous decades the major, often over-riding focus on technology selection and plant location was the availability of feed and process economics. Regarding feed availability, a desalination plant would often be located as close as possible to the source water to be desalinated. In the case of seawater desalination, it would be as close as feasibly possible to the coast and the seawater intake structures, minimizing the length of piping and pumping energy. Likewise for process economics, the particular technology selected would be based on the lowest cost of water production, taking into account the initial equipment and plant cost, the operating costs, and evaluating competing technologies using an economic framework such as net present benefit.

While this approach was largely acceptable within the social, regulatory and economic frameworks of the day, a major shift has occurred in recent times that is causing a major reconsideration of desalination technologies, in particular regarding the impact of the concentrate on the discharge environment. Social attitudes towards the discharge of concentrate to oceans (in the case of seawater desalination plants) and surface or sub-surface waters are hardening, which ultimately leads to greater challenges to licensing and even permission to operate. For desalination to continue to advance as a viable and acceptable method of producing water, adequate measures for concentrate management that are both economical and environmentally responsible need to be developed.

Very few texts have examined in detail the methods available for the management of concentrate, particularly in light of the latest desalination technologies and likewise the recent development of concentrate management strategies and technologies.

This book examines five methods used for concentrate management, namely; disposal to surface water, disposal to sewerage, deep well injection, land applications and evaporation ponds. In particular, the book focuses on the design, siting, cost, and environmental impacts of these methods. While these methods are widely practiced in a variety of settings already, there are many limitations that restrict the use of certain disposal options in particular locations.

References

Economic and Social Commission for Western Asia: Role of Desalination in Addressing Water Scarcity 2009. United Nations, New York (2009)

Fane, A.G., Wang, R., Jia, Y.: Membrane Technology: Past, Present and Future. In: Wang, L.K., Chen, J.P., Hung, Y.-T., Shammas, N.K. (eds.) Membrane and Desalination Technologies. Springer, New York (2011)

Khawaji, A., Kutubkhanah, I., Wie, J.: Advances in seawater desalination technologies. Desalination **221**(1–3), 47–69 (2008)

Chapter 2
Characteristics of Membrane Concentrate

Abstract This chapter discusses the characteristics of membrane concentrate, and the relevance that the concentrate has on the method of disposal. Membrane concentrate from a desalination plant can be regarded as a waste stream, as it is of little or no commercial benefit, and it must be managed and disposed of in an appropriate way. It is largely free from toxic components, and its composition is almost identical to that of the feed water but in a concentrated form. The concentration will depend on the type of desalination technology that is used, and the extent to which fresh water is extracted from the brine. Based on the treatment processes that are used, a number of chemicals may also be present in the concentrate, albeit in relatively small quantities.

Keywords Additional discharge streams · Antiscalants · Chemical treatment · Coagulation/flocculation · Concentration factor · Dechlorination · Filtration · High recovery · Membrane cleaning · Pretreatment · Recovery rate

2.1 Source Water

The concentrations of seawater and brackish water can vary significantly, and as such there is a difference between the concentrate produced from seawater desalination plants and brackish water desalination plants. Seawater typically has a level of total dissolved solids (TDS) between 33,000–37,000 mg/L. The average major ion concentration of seawater is shown in Table 2.1 along with water from the Mediterranean Sea, and water from Wonthaggi off the southern coast of Australia. Seawater salinity increases in areas where water evaporates or freezes, and it decreases due to rain, river runoff, and melting ice. The areas of greatest salinity occur and latitudes of 30° N and S where there are high evaporation rates,

B. Ladewig and B. Asquith, *Desalination Concentrate Management*,
SpringerBriefs in Green Chemistry for Sustainability,
DOI: 10.1007/978-3-642-24852-8_2, © The Author(s) 2012

Table 2.1 Major ion concentrations of seawater (mg/L)

Ion	Worldwide average	Mediterranean Sea[a]	Wonthaggi, Australia
Chloride, Cl^-	18,980	21,000–23,000	20,200
Sodium, Na^+	10,556	10,945–12,000	11,430
Sulphate, SO_4^{2-}	2,649	2,400–2,965	2,910
Magnesium, Mg^{2+}	1,272	1,371–1,550	1,400
Calcium, Ca^{2+}	400	440–670	420
Potassium, K^+	380	410–620	490
Bicarbonate, HCO_3^-	140	120–161	NR
Bromide, Br^-	65	45–69	62
Borate, $H_2BO_3^-$	26	NR	NR
Strontium, Sr^{2+}	13	5–7.5	7.6
Fluoride, F^-	1	1.2–1.55	0.9
TDS	34,482	38,000–40,000	NR

Data from Department of Sustainability and Environment 2008, Gaid and Treal 2007, Suckow et al. 1995

[a] Values for the Mediterranean Sea are taken from Gibraltar and Toulon (France)

Table 2.2 Concentration (mg/L) of major cations and anions in the brackish water feed to three desalination plants in the Abu Dhabi Emirate, United Arab Emirates

Ion	Al Wagan	Al Qua'a	Um Al-Zumool
Sodium, Na^+	741.59	451.13	2,482
Calcium, Ca^+	146.31	162.36	456.40
Magnesium, Mg^+	112	104	194
Potassium, K^+	28.46	27.24	110.1
Chloride, Cl^-	3,827	6,213	9,443
Phosphide, P^{3-}	–	0.14	–
Nitrate, NO_3^-	8.99	1.57	12.70
Sulphate, SO_4^{2-}	539.22	394.38	1746

Data from (Mohamed et al. 2005)

while there is a decrease in average salinity at the equator where there is a greater amount of precipitation (Millero et al. 2006). Areas of notably high salinity include the Mediterranean Sea and the Red Sea, which in parts can have concentrations of up to 39,000 and 41,000 mg/L respectively (Millero et al. 2006).

The concentration of brackish water is far more variable than seawater. Depending on the location, brackish groundwater can have TDS concentrations ranging from several hundred up to several thousand mg/L. Table 2.2 shows the ion concentrations from the feed of three brackish water desalination plants in the Emirate of Abu Dhabi, United Arab Emirates. It can be seen that even three water sources in the same region can have widely variable concentrations of dissolved salts.

Other components likely to be found in both seawater and brackish water are heavy metals and organic matter, both of which will impact the composition of the concentrate. Heavy metals and other toxic constituents are found in all water sources, but they are more likely to found in inland water sources rather than seawater (Del Bene et al. 1994). Metals may also be in the discharge as a result of corrosion. However, while thermal desalination plants are more susceptible to corrosion due to their operation at elevated temperatures, corrosion is not a significant problem in membrane desalination plants. Natural organic matter (NOM) will also be found in source waters, including humic acids and materials generated by algae or other organisms in the source water.

2.2 Recovery Rate and Concentrate TDS

The recovery of the feed water will determine how concentrated the final solution is. The recovery ratio, R, is defined as:

$$R = \frac{Q_P}{Q_F} = \frac{Q_P}{Q_C + Q_F} \tag{2.1}$$

where Q_P is the product flow rate, Q_F is the feed flow rate and Q_C is the concentrate flow rate. The recovery rate changes according to the specific membrane process, and in particular the number of membrane elements in each vessel, the number of membrane passes and stages per pass, the type of membrane that is used, and the quality of the final permeate (Mauguin and Corsin 2005). With lower levels of TDS, brackish water systems are able to achieve higher recovery rates, as the water is able to be concentrated to higher levels before the onset of precipitation, which eventually leads to membrane fouling. Brackish water desalination plants can achieve recoveries of between 65–90%, while seawater desalination plants range from 40–65% (Voutchkov et al. 2010).

Once the recovery rate is known, the TDS of the concentrate can be calculated with the following:

$$TDS_{concentrate} = TDS_{feed}\left(\frac{1}{1-R}\right) - \frac{R \times TDS_{permeate}}{100(1-R)} \tag{2.2}$$

Since the value of $TDS_{permeate}$ is often very low, it is often approximated as zero, and the above equation becomes:

$$TDS_{concentrate} = TDS_{feed}\left(\frac{1}{1-R}\right) \tag{2.3}$$

The concentration factor of the brine, the ratio of the concentrate and feed concentrations, can also be calculated. It is given as:

$$CF = \frac{TDS_{concentrate}}{TDS_{feed}} = \left(\frac{1}{1-R}\right) \tag{2.4}$$

For example, a plant operating at a recovery ratio of 90% will have a concentration factor of 10 (i.e., the components in the feed solution are concentrated 10 times).

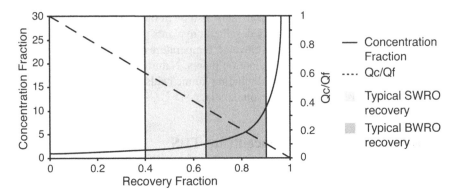

Fig. 2.1 Increase in concentration factor and decrease in concentrate volume with increasing recovery rate. The shaded regions represent typical recovery ranges for typical seawater reverse osmosis (SWRO) and brackish water reverse osmosis (BWRO) processes

The more water that is extracted, the smaller the volume of concentrate, but the greater the salinity of the brine. This trade off between concentration factor and recovery rate is shown in Fig. 2.1. As recovery approaches 100%, there is a sharp increase in the concentration factor, however the volume of the concentrate relative to the volume of feed, Q_c/Q_f is shown to decrease linearly. Figure 2.1 provides a useful comparison between the degree of concentration that may be achieved, and the volume of concentrate that must be managed. Higher recoveries are generally more desirable, as more water is recovered and less feed water is required to produce the same amount of water.

In terms of concentrate management, some disposal options are better suited to low volume and high concentration, while others are better suited to high volume with lower concentration. Ultimately there may be a balance between plant performance and disposal suitability. Plants with high recovery rates that produce a more concentrated brine must have the means to dispose of this concentrate in a safe and environmentally friendly manner. When suitable concentrate management is not achievable, the recovery of a plant may need to be compromised.

Additional processing of concentrate to achieve very high recoveries, that eventually approach 100% recovery and zero liquid discharge is possible. The definition of high recovery processing is variable, but it can be generally described as further processing of concentrate to achieve a higher recovery than what would normally be achieved in a single membrane pass. For brackish water desalination, this generally translates to recovery rates of greater than 90%. Many different methods exist for the treatment of brine to achieve high recovery. These involve some form of control over the precipitation of sparingly soluble salts and other potential foulants, so as to be able to remove them from the system at a convenient stage and to allow further treatment.

Table 2.3 Calculated and actual ion concentrations (mg/L) in the feed and concentrate from the Al Wagan BWRO desalination plant

Ion	Feed	Calculated		Actual	
		Concentrate	Concentration factor	Concentrate	Concentration factor
Sodium, Na^+	742	2,473	3.33	2,248	3.03
Calcium, Ca^+	146	487	3.33	367.96	2.51
Magnesium, Mg^+	112	373	3.33	282.02	2.52
Potassium, K^+	28	93	3.33	68.44	2.40
Chloride, Cl^-	3827	12757	3.33	8,946	2.34
Nitrate, NO_3^-	9	30	3.33	7.11	0.79
Sulphate, SO_4^{2-}	539	1,797	3.33	1,540	2.86

Data from (Mohamed et al. 2005)

2.2.1 Case Example

Data from the Al Wagan BWRO desalination plant (Mohamed et al. 2005) has been used to highlight a typical brackish water desalination scenario. The following has been calculated assuming 200 m³/day of brackish water is treated at a recovery of 70%. From Eq. 2.1:

$$0.70 = \frac{Q_P}{200}$$

$$Q_P = 140 \ m^3/d$$

Therefore the permeate quantity is 140 m³/d. The volume of concentrate can then be determined using $Q_F = Q_P + Q_C$. With a feed of 200 m³/day and a recovery of 70%, the concentrate quantity is 60 m³/day. The concentration factor can be found using Eq. 2.4:

$$CF = \left(\frac{1}{1-R}\right) = \left(\frac{1}{1-.7}\right)$$

$$CF = 3.33$$

Table 2.3 shows the calculated ion concentrations with a concentration factor of 3.33, and compares them with actual plant values. Note that there are significant differences between the calculated and actual ion concentrations for most species, however this is likely to be a result of the plant operating below the stated recovery rate of 70%. The differences in the concentration factors from the actual plant data (ranging from 0.79 for nitrate to 3.03 for sodium) are largely a result of the different precipitation points for each of the species in the water.

2.3 Chemical Treatment and Additional Discharge Streams

The majority of the discharge from a desalination processes is concentrated brine from the membrane process, and this may contain quantities of treatment chemicals used. Treatment of water is necessary in all desalination plans for variety of reasons; feed water treatment, membrane protection, membrane cleaning, permeate treatment and concentrate treatment prior to discharge. Although non-chemical treatment is possible, chemical treatment is widely practiced.

In addition to the concentrated brine, streams from the pretreatment, cleaning, post-treatment and startup processes are also present. The management of these streams, including chemical treatment, must be tailored to suit the specific concentrate management scheme in place. Buffer tanks can be used to store additional discharge streams, thus helping to regulate the flow and concentrate of the final discharge (Mauguin and Corsin 2005). Streams which cannot be disposed of via the same means as the concentrate will require alternative treatment and disposal.

2.3.1 Pretreatment

Before any membrane treatment process, the feed water must undergo pretreatment. This is primarily done to reduce fouling caused by suspended solids, microbial growth, and inorganic deposits of silica and precipitated sparingly soluble salts and silica. Fouling reduces the flux of water through the membranes and hence lowers the water recovery of the process. Excessive fouling can incur additional costs for the replacement of membranes. Pretreatment is also used to prevent the oxidation and hydrolysis of the membranes, and remove metals or other chemicals in the feed water as necessary. Typical pretreatment for a reverse osmosis process includes initial screening, or biofouling control (often referred to as chlorination), acid treatment, coagulation/flocculation, media filtration, antiscalant addition, cartridge filtration and dechlorination (Isaias 2001). A summary of the chemicals that are used for these processes can be seen in Table 2.4. Alternatively membrane pretreatment can be used, reducing the amount of chemicals that are required.

2.3.1.1 Biofouling Control and Dechlorination

Biofouling control is used to prevent the buildup of microorganisms and the formation of biofilms that cause fouling. The oxidants chlorine and ozone can be added as biocides, with chlorine being used most commonly. Chlorine is added in the form of chlorine gas (Cl_2) or sodium hypochlorite (NaOCl) (Fritzmann et al. 2007). The dosage of the oxidant must be high enough to allow for residual oxidant to disinfect the entire pre-treatment processes (Greenlee et al. 2009).

Table 2.4 Commonly used membrane desalination plant chemicals

Pretreatment process	Typical chemicals
Biofouling control	Chlorine, sodium hypochlorite, calcium hypochlorite, ozone
Antiscalants	Polymeric substances such as polyphosphates, phosphonates and polycarbonic acids
Acids	Sulphuric acid, hydrochloric acid
Coagulants/ Flocculants	Ferric chloride, ferric sulphate, polyelectrolytes
Dechlorination	Sodium bisulphite

Data from (Voutchkov et al. 2010; Committee on Advancing Desalination Technology, National Research Council 2008; Younos 2005)

All oxidants used must be removed in the final stage of the pretreatment process, as they are known to damage most polymer membranes used for desalination. In particular, chlorine is known to be harmful to commonly used thin-film composite polyamide membranes.

When chlorine is used as an oxidant, sodium bisulphite can be used for dechlorination. However, even after the process of dechlorination, free residual chlorine (FRC) may be present in the discharge. The sodium bisulphite used for dechlorination may also cause low levels of dissolved oxygen in the concentrate. For processes which use ozone, not only must it be removed to prevent damage to oxidant sensitive membranes, but also to prevent the formation of bromate, a known carcinogen, in waters containing bromide (Greenlee et al. 2009).

As an alternative to the addition of oxidants, ultraviolet (UV) treatment may be used. UV treatment has found to be less effective than chlorine or ozone in preventing biofouling, however unlike those methods, it does not incur additional downstream processing (Committee on Advancing Desalination Technology, National Research Council 2008; Cotruvo 2005). This eliminates the need for the use sodium bisulphite or other reducing agents and the potential for FRC or ozone to be present in the final concentrate.

2.3.1.2 Acid treatment

To increase the solubility of calcium carbonate and reduce its potential for precipitation and hence membrane fouling, the pH of the feed water is lowered through the addition of acid (Greenlee et al. 2009). Sulphuric acid and hydrochloric acid are most commonly used. Acid treatment can also help to improve the coagulation of colloids and slightly increase the solubility of silica (Bergman 2007).

2.3.1.3 Coagulation/Flocculation

Coagulants are used to bind together particulate and colloidal matter so they may be filtered from the feed before the membrane process. Coagulants can be either inorganic (such as ferric salts) or organic polyelectrolytes. The correct dosage and effective removal of coagulants is crucial for the prevention of membrane fouling (Bergman 2007). These chemicals are only likely to be present in the concentrate if the filter backwash is disposed of with the concentrate and does not undergo any additional treatment (Lattemann and Höpner 2008).

2.3.1.4 Antiscalants

The extent of process recovery is often limited by the fouling of membranes from sparing soluble precipitates. Antiscalants are added during pretreatment to increase the solubility of salts likely to precipitate, enabling the membrane process to achieve a higher recovery before fouling occurs. Antiscalants can be a number of polymeric substances (typically polyphosphates, phosphonates and polycarbonic acids), and as there is no treatment process to remove antiscalant, they will be present in the membrane concentrate discharge.

2.3.1.5 Filtration

Both media filters and cartridge filters can be used in a pretreatment process. Granular media filters involve the filtration of large particles through different layers of fine particles, usually coal, pumice, sand or garnet (Bonnelye et al. 2004). Cartridge filters act as the final filtration step before the water passes through the membranes, and remove fine particles as small as 1μm.

Although no chemicals are required to pass water through the filter, the filter cleaning process produces a stream of waste known as filter backwash. This discharge is high in suspended solids, but the concentration of dissolved salts is similar to the intake. It will also contain any coagulants or flocculants that were added to the water. This water generally requires treatment prior to disposal with the membrane concentrate (Mauguin and Corsin 2005). Treatment of filter backwash involves the separation of solids from the waste stream to bring the water quality back to a similar quality of feed water. The separated solid waste can be then separated and disposed of appropriately.

2.3.1.6 Membrane Pretreatment

Microfiltration (MF) and ultrafiltration (UF) membranes can be used as forms of pretreatment for nanofiltration (NF) or reverse osmosis (RO) desalination processes. Membrane pretreatment reduces the amount of chemicals that are required and hence reduces the environmental impact of the final discharge. MF membranes can be used to filter particles with diameters of 0.1–10 μmm and typically remove bacteria, viruses, precipitates, coagulates and large colloidal particles. UF can remove particles with diameters as small as 0.002 μm, and

Table 2.5 Commonly used cleaning chemicals for the removal of membrane foulants

Foulant	Typical cleaning chemicals
Metal oxides	Citric acid, phosphoric acid
Inorganic colloids	Sodium hydroxide, sodium dodecyl sulphate, sodium dodecyl benzene sulphonate
Biofilms	Sodium ethylenediaminetetraaceticacid (Na-EDTA), sodium triphosphate, trisodium phosphate
Organic matter	Sodium hydroxide, sodium dodecyl sulphate, sodium dodecyl benzene sulphonate
Silica	Ammonium bifluoride, sodium hydroxide, sodium dodecyl benzene sulphonate

Data from (Committee on Advancing Desalination Technology, National Research Council 2008; Lattemann S and Höpner 2003)

typically remove high molecular weight proteins, large organic molecules and pyrogens (Curcio 2009).

Membrane pretreatment can improve the feed water quality and reduce particulate matter to a greater degree than other pretreatment methods. This reduces the required cleaning frequency of the downstream membranes and the amount of cleaning chemicals that are required and subsequently disposed of (Pearce 2007).

Similar to filter backwash, the concentrate from these membranes requires treatment before it can be disposed of with the membrane concentrate. However, the total amount of solids produced after the treatment of filter backwash can be 60–80% greater than MF and UF concentrate due to the addition of coagulants prior to the granular media filters (Bergman 2007).

NF has been also been studied as a potential form of pretreatment for reverse osmosis desalination processes (Hassan et al. 1998, 2000). Based on the feed water, it may be a suitable pretreatment method that allows for operation with little or even no use of antiscalants.

2.3.2 Membrane Cleaning

Cleaning chemicals are used to remove membrane fouling. Fouling can include salt precipitation, particulate or colloidal fouling, organic fouling and biofouling. Both acidic and alkaline solution are used to clean membranes based on the foulants that accumulate on the membrane surface. A list of common membrane cleaning chemicals for foulant removal is provided in Table 2.5. Membranes are also disinfected with chemicals such as sodium hypochlorite or hydrogen peroxide (Van Der Bruggen et al. 2003).

Cleaning chemicals are generally discharged in a separate waste stream rather than the concentrate discharge, as additional treatment is often required.

However, because both acids and bases are used in cleaning, it may be possible to store these chemicals in a stirred buffer tank to allow the solution to neutralise. This solution may then may be slowly added to the concentrate waste to dilute it and allow for a safe disposal (Mauguin and Corsin 2005). Note that the volume of this discharge is much lower than both the concentrate and filter backwash discharges.

2.3.3 Other Discharges

While the plant is in operation, but not producing water that is of the required quality, two additional discharge streams can be produced (Mauguin and Corsin 2005). This usually occurs during the startup of the plant. The first stream is the pretreated water which has not reached an acceptable quality to pass through the membrane unit. Its composition is largely similar to that of the feed, and if membrane processes are not used as pretreatment, its salinity should be the same as feed. The second stream is the permeate which has not yet reached the desired quality of the final water product. This stream will have a lower salinity than the feed, and should not contain great quantities of harmful chemicals. Both these streams should be safe for disposal in the concentrate stream.

References

Bergman, R.: Reverse Osmosis and Nanofiltration. American Water Works Association Denver (2007)

Bonnelye, V., Sanz, M.A., Durand, J.-P., Plasse, L., Gueguen, F., Mazounie, P.: Reverse osmosis on open intake seawater: pre-treatment strategy. Desalination **167**, 191–200 (2004)

Committee on Advancing Desalination Technology, National Research Council: Desalination: A National Perspective. The National Academies Press, Washington, D.C. (2008)

Cotruvo, J.A.: Water Desalination Processes and Associated Health and Environmental Issues. Water Cond. Purif. (January), 13–17 (2005).

Curcio, E., Drioli, E.: Membranes for Desalination. In: Cipollina, A., Micale, G., Rizzuti, L. (eds.) Seawater Desalination: Conventional and Renewable Energy Processes. pp. 41–75. Springer, Heidelberg; New York (2009)

Del Bene, J.V., Jirka, G., Largier, J.: Ocean Brine Disposal. Desalination **97**(1–3), 365–372 (1994).

Department of Sustainability and Environment: Victorian Desalination Project Environmental Effects Statement—Volume 2. (2008)

Fritzmann, C., Löwenberg, J., Wintgens, T., Melin, T.: State-of-the-art of reverse osmosis desalination. Desalination **216**(1–3), 1–76 (2007).

Gaid, K., Treal, Y.: Le dessalement des eaux par osmose inverse: l'expérience de Véolia Water. Desalination **203**(1–3), 1–14 (2007). doi:10.1016/j.desal.2006.03.523

Greenlee, L.F., Lawler, D.F., Freeman, B.D., Marrot, B., Moulin, P.: Reverse osmosis desalination: Water sources, technology, and today's challenges. Water Res. **43**(9), 2317–2348 (2009).

Hassan, A.M., Al-Sofi, M.A.K., Al-Amoudi, A.S., Jamaluddin, A.T.M., Farooque, A.M., Rowaili, A., Dalvi, A.G.I., Kither, N.M., Mustafa, G.M., Al-Tisan, I.A.R.: A new approach to membrane and thermal seawater desalination processes using nanofiltration membranes (Part 1). Desalination **118**, 35–51 (1998).

Hassan, A.M., Farooque, A.M., Jamaluddin, A.T.M., Al-Amoudi, A.S., Al-Sofi, M.A.K., Al-Rubaian, A.F., Kither, N.M., Al-Tisan, I.A.R., Rowaili, A.: A demonstration plant based on the new NF-SWRO process. Desalination **131**, 157–171 (2000).

Isaias, N.P.: Experience in reverse osmosis pretreatment. Desalination **139**(1–3), 57–64 (2001). doi:10.1016/s0011-9164(01)00294-6

Lattemann, S., Höpner, T.: Environmental impact and impact assessment of seawater desalination. Desalination **220**(1–3), 1–15 (2008).

Lattemann, S., Höpner, T.: Seawater Desalination: Impacts of Brine and Chemical Discharge on the Marine Environment. Desalination Publications, L'Aquila (2003)

Mauguin, G., Corsin, P.: Concentrate and other waste disposals from SWRO plants: characterization and reduction of their environmental impact. Desalination **182**(1–3), 355–364 (2005).

Millero, F.J.: Chemical Oceanography, Third Edition. CRC Press, Boca Raton (2006)

Mohamed, A.M.O., Maraqa, M., Al Handhaly, J.: Impact of land disposal of reject brine from desalination plants on soil and groundwater. Desalination **182**(1–3), 411–433 (2005).

Pearce, G.K.: The case for UF/MF pretreatment to RO in seawater applications. Desalination **203**, 286–295 (2007).

Suckow, M.A., Weisbroth, S.H., Franklin, C.L.: Salinity in the oceans. In: Mark, A.S., Steven, H.W., Craig, L.F. (eds.) Seawater (Second Edition). pp. 29–38. Butterworth-Heinemann, Oxford (1995)

Van Der Bruggen, B., Lejon, L., Vanecasteele, C.: Reuse, Treatment, and Discharge of the Concentrate of Pressure-Driven Membrane Processes. Environ. Sci. Technol. **37**(17), 3733–3738 (2003).

Voutchkov, N., Sommariva, C., Pankratz, T., Tonner, J.: Desalination Process Technology. In: Cotruvo, J.A., Voutchkov, N., Fawell, J., Payment, P., Cunliffe, D., Lattemann, S. (eds.) Desalination Technology—Health and Environmental Impacts. CRC Press, Boca Raton (2010)

Younos, T.: Environmental Issues of Desalination. J. Contemp. Water Res. Edu. **132**(1), 11–18 (2005).

Chapter 3
Process Feasibility

Abstract The feasibility of a method of concentrate management depends on a wide range of factors. The selection is very much site specific, and for each site often only one or two concentrate management options are feasible. Of particular importance is the quality of the concentrate, the cost of the process, any potential environmental impacts and regulations surrounding the method of concentrate management.

Keywords Costs · Disposal regulations · Environmental impacts · Process selection

3.1 Process Selection

The following general factors must be considered when determining the most suitable method of concentrate management (Younos 2005):

- Concentrate volume
- Concentrate quality and composition
- Location
- Cost
- Environmental regulations
- Public acceptance
- Future plant expansion

The concentrate volume and plant location will tend to be the two major factors that determine the suitability of a particular option. The concentration of the brine is also important, as the severity of potential environmental impacts generally increases with increasing concentration. The siting of a desalination plant must take into account the availability of disposal options, along with potential sources of feed water for the plant and the proximity to the end user.

B. Ladewig and B. Asquith, *Desalination Concentrate Management*,
SpringerBriefs in Green Chemistry for Sustainability,
DOI: 10.1007/978-3-642-24852-8_3, © The Author(s) 2012

Table 3.1 General cost factors for conventional methods of disposal. Data from Younos (2005)

Cost factor	Disposal method			
	Surface water	Deep well injection	Spray irrigation	Evaporation pond
Pipes and pumps	X	X	X	X
Treatment system	X	X	X	–
Outfall structure	X	–	–	–
Injection well construction	–	X	–	–
Monitoring Wells	–	X	X	X
Land, land preparation	–	–	X	X
Wet weather storage	–	–	X	–
Alternative disposal	–	X	–	–
Subsurface drainage	–	–	X	–

3.2 General Concentrate Management Costs

While the cost of membrane based desalination has decreased over time with improving process efficiency, the cost of concentration disposal has remained relatively constant. Furthermore, disposal cost is unlikely to decrease in the future due to the simplistic and low-tech nature of the equipment required for concentrate management, and the range of nontechnical factors and limitations that determine the feasibility for each option (Mickley 2009).

Each desalination plant will have greatly different disposal costs based on the methods available. A summary of the range of factors which contribute to these costs can be seen in Table 3.1. Note that the only common costing factor across all methods of management is the cost of transporting concentrate to the final disposal site. This cost increases with increasing distance between the disposal site and the plant, so it becomes advantageous to site a desalination plant as close as practicable to the site of disposal. For plants that utilise deep well injection, the cost of pumping may be significantly more than other methods due to the high injection pressures that are often required. More detailed discussions on these cost factors can be found in Chaps. 4, 5, 6 and 7 of this book.

The relative increase in cost for surface water disposal, deep well injection, spray irrigation (land applications) and evaporation ponds can be seen in Fig. 3.1. It can be seen that surface water disposal is the cheapest alternative, and it has a strong economy of scale as concentrate volume increases. Deep well injection also has a strong economy of scale, but this method's high construction costs means that it is only becomes feasible with a high enough disposal volume. Evaporation ponds have a poor economy of scale, and it can be seen that the overall cost increases rapidly with volume. This is due to the large amounts of land that are required as the volume of concentrate increases.

Fig. 3.1 Relative capital costs of common methods for concentrate disposal with increasing concentrate flow rate. Adapted from Mickley (2009)

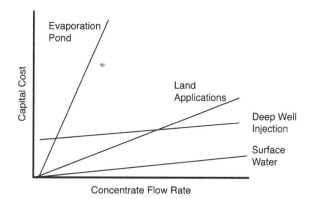

3.3 Regulations

Before any concentrate management scheme can be implemented, local regulations must be met and any necessary permits acquired. Due to the environmental impacts of the constituents in the concentrate, regulations are often based on the concentration of solution to be disposed rather than the volume (Ahmed et al. 2000). Disposal regulations and permits are defined by various governing bodies, which differ according to the location of the plant and the desired site of disposal.

3.4 Environmental Impacts

The siting, design, construction and management of any disposal option needs to be done to minimise negative environmental impacts. In some instances, there may be a trade off between desalination plant performance and the environmental suitability of the brine. Plants with high recovery rates that produce a more concentrated brine must have the means to dispose of this concentrate solution in a safe and environmentally friendly manner. When safe concentrate management is not achievable, the recovery of a plant may need to be compromised.

High recovery is generally considered as beneficial, as the lower volume of concentrate that is produced is easier to deal with and requires less cost for transport.

References

Ahmed, M., Shayya, W.H., Hoey, D., Mahendran, A., Morris, R., Al-Handaly, J.: Use of evaporation ponds for brine disposal in desalination plants. Desalination **130**(2), 155–168 (2000)

Mickley, M.: Treatment of Concentrate, Desalination and Water Purification Research and Development Program Report No. 155. U.S. Department of the Interior, Bureau of Reclamation, Denver (2009)

Younos, T.: Environmental Issues of Desalination. J. Contemp. Water Res. Edu. **132**(1), 11–18 (2005)

Younos, T.: The Economics of Desalination. J. Contemp. Water Res. Edu. **132**(1), 39–45 (2005)

Chapter 4
Disposal to Surface Water

Abstract This chapter discusses disposal to surface water, the most common method of concentrate management. This includes concentrate that is directly disposed of into rivers, creeks, lakes, oceans, bays, and other bodies of water. Concentrate is piped to the site of disposal, where it is discharged to the receiving body of water via an outfall structure. The environmental impacts of surface water disposal may be lessened by diluting the concentrate prior to discharge, or by dilution of the concentrate through the design of the outfall structure and diffusers. Pretreatment processes that lessen the impact on the environment should also be considered.

Keywords Concentrate blending · Concentrate transport · Costs · Disposal regulations · Environmental concerns · Outfall design

4.1 Site Selection

The availability of a body of water for disposal is a major consideration in the siting of desalination plant. Even with a suitable body of water, the outfall needs to be designed and located to reduce environmental impact. The extent of mixing and dilution based on the properties of the concentrate and the receiving water must therefore be considered. The behavior of concentrate following discharge is determined by its salinity, density and flow rate, as well as the hydrography and biological properties of the receiving body of water. These include depth, volume, temperature, water composition and additionally for ocean outfalls, tides, waves and currents (Lattemann and Höpner 2008). Due to the variation in these conditions, each receiving body of water will be unique, and so the specific conditions and impacts must be determined on a case-by-case basis before any discharge can occur.

B. Ladewig and B. Asquith, *Desalination Concentrate Management*,
SpringerBriefs in Green Chemistry for Sustainability,
DOI: 10.1007/978-3-642-24852-8_4, © The Author(s) 2012

Perhaps the most pertinent characteristic that will determine its mixing and dilution is its density. The density of the concentrate varies with salinity; concentrate with a salinity higher than the receiving water will be negatively buoyant, and that with a lower salinity will be positively buoyant. Unless diluted prior to discharge, membrane concentrate will have a greater salinity and hence greater density than the receiving water. This results in a plume which sinks upon discharge.

Seawater desalination plants almost exclusively use surface water disposal, and they are often sited so that concentration may be discharged directly back to the ocean with minimal environmental impact. It is recommended that ocean discharges be located along open coast, as opposed to locations where there will be minimal water movement, such as estuaries (Watson et al. 2003). Ideal ocean bottom profiles are ones which achieve a sufficient depth quickly, as this reduces the require length of the outfall pipe and decreases costs (Mauguin and Corsin 2005).

Inland desalination plants are much more limited in the availability of a suitable discharge site. Inland water bodies are often high in quality and have potential for use as water sources, limiting their suitability. Disposal is only feasible if the quality of the concentrate is high enough to be compatible with the receiving body of water (Younos 2005).

If the concentrate is discharged to the same body of water as the feed water source, both the intake and outfall should be located so as to not to interfere with one another. The presence of any other local desalination or wastewater discharge is an important factor, as these may increase the salinity of the receiving water or reduce its quality, thereby reduce the compatibility of the concentrate.

4.2 Design

The two major design factors for surface water discharge are the transport of concentrate to the outfall site, and the design of the outfall structure. The design of the transport system and the outfall structure will depend upon the proximity of the plant to the site of disposal, and the body of water into which the concentrate will be discharged.

4.2.1 Concentrate Transport

Concentrate must be transported from the desalination plant to the site of disposal, typically done via above or underground pipework. Tunneling and the installation of underground pipework is more costly than aboveground pipework, however this can be necessary when the outfall is located underwater, or when the plant is located some distance from the discharge point. The pipes used to transport concentrate should be resistant to corrosion. This can be done by constructing them from corrosion resistant steels with coatings to prevent oxidation, utilising cathodic protection, or using plastic instead of steel (Bergman 2007). When there are no other feasible concentrate disposal options, some inland sites may be required to haul brine to an ocean site for disposal.

4.2.2 Outfall Structure

At the site of discharge, the concentrate is released into the water via an outfall. A well-designed outfall structure should be effective in promoting mixing and helping minimize potential environmental impacts. The variety of outfalls range from simple discharge from the end of a pipe, to more complex structures involving long multi-port diffusers. The specific design of the outfall will be based on the expected mixing and dilution of the brine, and its compatibility with the receiving water. This particularly refers to the density and buoyancy of the concentrate.

Unless the body of water has a high flow rate or high turbulence that allows for sufficient dilution, simple discharge from the end of a pipe should be avoided. This type of simple outfall is more suited to small discharges into ocean tidal zones where large amounts of energy allow for rapid dilution and mixing (Voutchkov et al. 2010). Larger concentrate volumes should avoid discharge into tidal zones to prevent the accumulation of salt in these areas (Voutchkov et al. 2010). The use of diffusers is far more common, which is a pipe or series of pipes with numerous discharge ports. These can be designed to meet mixing and dilution requirements based on the concentrate and the receiving water. A longer pipe with multiple ports increases dilution by spreading the discharge out over a wider area. To maintain a constant discharge flow along the pipe and to allow for a better dilution, the size of the discharge ports increases along the pipe as the pressure head decreases.

Dilution can be promoted by the velocity of the discharge at each port, and the angle of discharge. Figure 4.1 shows the general geometry of a jet of dense concentrate discharging into a body of water. The optimal angle and velocity of the discharge will vary with the outfall depth, the incline of the sea bed, and the local currents and tides. Discharge velocity, and hence the level of dilution that can be achieved, is largely influenced by the nozzle diameter. Smaller nozzles help to promote greater dilution, and increase the vertical and horizontal distance of the jet (Cipollina et al. 2004). Currents and tides can also improve the mixing of concentrate if the discharge is orientated correctly.

The modeling and optimisation of concentrate discharge to reduce environmental impacts has been the focus of numerous studies (Al-Barwani and Purnama 2007, 2008; Alameddine and El-Fadel 2007; Bleninger and Jirka 2008; Malcangio and Petrillo 2010; Purnama and Al-Barwani 2004; Purnama et al. 2003). Much of this research has been for a project by the Middle Eastern Desalination Research Center, Environmental Planning, Prediction and Management of Brine Discharges from Desalination Plants.

Numerous software packages can be used to assist in the design and optimisation of outfall structures. *CORMIX* is a software modeling package designed to

Fig. 4.1 A negatively
buoyant discharge jet
releasing concentrate
into a body of water
with a inclined bed

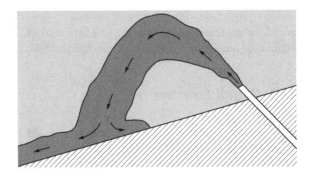

Fig. 4.2 Typical duckbill
valve in a closed position that
may be used on an outfall to
prevent backflow

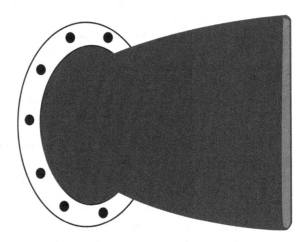

model mixing zones of brine discharges and is a useful tool for environmental
impact assessments (Mixon Inc 2011). Additionally, *Plumes* has been developed
by the American EPA to primarily model wastewater discharge, however it can
also be applied to concentrate discharge (United States Environmental Protection
Agency 2011).

To prevent backflow and to prohibit organisms entering the pipe, duckbill
valves can be installed on the end of diffuser ports. An example of this can be seen
in Fig 4.2. These valves can be selected with varying operating pressures to ensure
all valves along a diffuser open at the same time as the pressure of the concentrate
drops along the pipe (Mauguin and Corsin 2005).

4.3 Costs

Discharge to surface water is the most economical form of concentrate management
for seawater desalination plants, regardless of the discharge volume. Due to the
availability of ocean discharge for seawater desalination plants, the cost of disposal
tends to be less costly than for inland desalination. Costs include pumps and pipes,

any pretreatment prior to discharge, the outfall structure, and the monitoring of the receiving water quality. In some instances, pumps may not be required if the pressure of the concentrate is high enough to transport it to the site of disposal.

Pipe costs increase with increasing volume and increasing distance between the desalination plant and the site of disposal. While discharge from the end of a pipe can be quite inexpensive, costs will begin to increase the further the outfall is from the shore, the deeper the outfall is, and the more complex the outfall structure design is. This is particularly due to dredging, trenching and tunneling, which can be three to four times more expensive underwater than on land (Mickley 2006). The cost of these civil works are approximately 3% of the total construction cost for a seawater reverse osmosis plant (Sommariva 2010). The cost of monitoring will vary based on the environmental sensitivity of the discharge area (Schliephake et al. 2005). The overall cost of disposal may be reduced through the simultaneous discharge of multiple effluent streams from different sites.

4.4 Environmental Concerns

Concentrate can be harmful to the environment due to either its higher than normal salinity, or due to pollutants that otherwise would not be present in the receiving body of water. These include chlorine and other biocides, heavy metals, antiscalants, coagulants and cleaning chemicals. Of particular concern is the effect of pollutants on delicate ecosystems and endangered or threatened species. However, with appropriate measures in place, the discharge of concentrate to surface water can remain a viable method for seawater desalination plants.

4.4.1 Increased Salinity

As a result of the high density and negative buoyancy of membrane concentrate, benthic communities and other non-mobile species are most affected by increases in ambient salinity. Most organisms will generally be able to adapt to minor changes in salinity, and while some species may adapt well to an increase in salinity, many will not be able to withstand long-term exposure to higher than normal concentrations (Lattemann and Höpner 2008). Unfortunately, little data is available regarding the tolerance of various species to changes in salinity. It has been estimated that benthic communities can withstand variations in salinity of ± 1000 ppm (Del Bene et al. 1994). For most marine organisms that live in seawater, the salinity tolerance can be up to and potentially even greater than 40% of the average annual ambient ocean salinity concentration (Voutchkov et al. 2010).

4.4.2 Pretreatment Chemicals

Any FRC that may be present in the concentrate is known to be toxic, and can have severe impacts on marine life. However, following dechlorination with sodium bisulfite, the level of free residual chlorine in the concentrate is often quite low, and quickly decreases after discharge as it dissipates and degrades (Lattemann and Höpner 2008). Chlorine also has potential to form halogenated compounds, and although these can be dangerous to marine life, their concentrations are often well below the FRC concentrations, and hence considered less toxic (Lattemann and Höpner 2008).

Antiscalants are found in small concentrations in desalination discharge. Polymer antiscalants are of low toxicity, and have little environmental impact (Lattemann and Höpner 2008). In aquatic environments, polymer antiscalants behave in a similar manner to humic substances, and they are unlikely to accumulate in aquatic life (Lattemann and Höpner 2003).

Coagulants, which are present in filter backwash, are of low toxicity and are not considered a major environmental concern. One of the greatest effects of coagulants comes through the use of ferric salts, which are likely to cause colouration and increase the turbidity of the backwash (Lattemann and Höpner 2008). The quantity of coagulants in the concentrate can in part be regulated by the slow release of filter backwash from a buffer tank into the concentrate flow.

4.4.3 Cleaning Chemicals

Most cleaning chemicals used for membrane desalination plants are harmful to the environment. Common cleaning chemicals are outlined in Table 2.5. These cleaning solutions should be treated prior to discharge with membrane concentrate, or sent to a location where appropriate treatment can occur.

Discharge of solutions which are either basic or acid are dangerous to marine life, and should be neutralised prior to discharge. This can be achieved with a buffer tank to store and neutralise acidic and alkaline solutions with each other. Seawater may also be used to neutralise cleaning solutions.

Compared with the volume of concentrate, the volume of cleaning solution is relatively small. To reduce environmental impacts, as well as costs associated with cleaning, the quantity of chemicals that are used can be reduced through an effective pre-treatment process.

4.4.4 Heavy Metals

Metal in the discharge can come from the source water, or be a product of corrosion. Note that due to their operation at elevated temperatures, when compared with membrane desalination plants, thermal desalination plants are more likely to have occurrences of pipe corrosion and hence metals present in the discharge.

Upon discharge, metals form organic and inorganic complexes, or absorb to reactive surfaces, and subsequently sink, leading to an accumulation in sediments (Höpner 1999). Due to this, benthic organisms may be more likely to assimilate metals from the discharge (Lattemann and Höpner 2008). Metals can have acute effects on marine organisms, however membrane concentrate has very low concentrations of metals, and so their environmental impact is usually minimal.

4.4.5 Dissolved Gases

Concentrate from membrane desalination plants may have low levels of dissolved oxygen, and high levels of hydrogen sulfide and carbon dioxide. Low levels of oxygen can be harmful to aerobic marine organisms. This may be a result of the characteristics of the feed water, or from the use of sodium bisulfite for dechlorination (Lattemann and Höpner 2003). Increasing salinity contributes slightly to the amount of dissolved oxygen in the concentrate, since the solubility of oxygen decreases as the salinity increases. High levels of hydrogen sulfide and carbon dioxide may also occur in the discharge as a result of the feed water characteristics. Low amounts of dissolved oxygen and high amounts of dissolved hydrogen sulfide and carbon dioxide are more likely to occur from brackish water sources than from seawater. Prior to discharge, the concentrate can undergo processes of aeration and degasification to increase dissolved oxygen content and to reduce the concentration of hydrogen sulfide and carbon dioxide.

4.5 Concentrate Blending

When the design of an outfall is unable to produce a sufficiently diluted mixing zone that is below the maximum permitted concentration, additional dilution prior to discharge may be necessary. Brine may be diluted prior to discharge by blending it with wastewater from treatment plants, power plants, or other water sources that may be available. Blending reduces the concentration of the discharge and brings its salinity closer to that of the receiving body of water, allowing for more rapid mixing and dilution. Further benefits include reduced capital costs of building additional tunnels and outfall structures, and the possibility of a modified discharge permit, rather than an application for a new one (Mickley 2009). It is important to note that dilution may not necessarily negate the effects of chemicals in the discharge, as the total load and the accumulation over time determines the environmental impact (Höpner 1999).

4.5.1 Blending with Sewage and Wastewater

In some instances, discharge is made directly to a sewer where it subsequently passes through a wastewater treatment plant. Due to the adverse effects that high salinity can have on these plants, this is generally only suitable with small flows of concentrate into relatively large treatment facilities. Due to its low salinity, concentrate from both microfiltration and ultrafiltration plants is often discharged to sewer (Schliephake et al. 2005).

Discharge to sewer will need to be approved by the downstream wastewater treatment plant. The viability of disposal is governed by the downstream treatment plant's capability to handle both the volume and concentrate of the brine, and the final concentration of the effluent from the facility. If the salinity of the concentrate is too high, it may hinder the biological treatment process (World Health Organisation 2007). Furthermore, if the salinity of the treatment plant's effluent becomes too high, environmental and regulatory problems may be encountered during final disposal. In this case, additional permits or alternative disposal solutions may be required for the wastewater plant. Desalination plants may also be required to pay a fee to the wastewater treatment plant.

Alternatively, if the desalination plant is located near an existing wastewater treatment plant, the concentrate may be blended with the treatment plant's effluent before final disposal to water. This process improves the mixing of the discharge with the receiving water by blending a high salinity, negatively buoyant stream with a low salinity, positively buoyant stream. In this instance, the existing wastewater treatment plant outfall must be able to handle the capacity of the combined discharge. Additional fees to the wastewater treatment plant may also be necessary. Unfortunately, this method of disposal is not widely practiced and has limited applications as a result of the necessary conditions, including the capacity limitations of the outfall, and the environmental impacts of the blended streams (Voutchkov et al. 2010).

Examples of desalination plants that currently blend their concentrate with treatment plant outfall include the Thames Water Desalination Plant in London (150,000 m^3/day capacity) and the Barcelona Seawater Desalination Plant (200,000 m^3/day capacity).

4.5.2 Blending with Power Plant Cooling Water

Co-location of a power plant and a seawater reverse osmosis desalination plant allows for the cooling water from a neighbouring power plant to be blended with the waste from a desalination plant before discharge (Voutchkov 2004). In such a process, seawater is used as the cooling water for the condensers in a power plant. This water is then used as both the feed for the desalination process, and for blending to dilute the concentrate from the desalination plant.

From a concentrate disposal point of view, a number of benefits exist from the co-location of a desalination plant and a power plant (Voutchkov 2004). These include the dilution of the concentrate to improve its compatibility with the receiving water, and cost reductions in sharing an outfall structure. When the concentration of the discharge is reduced, the environmental impacts of high saline discharge are minimised. The salinity of the discharge can be as low as the natural variation in seawater, and so this lower saline discharge requires a shorter outfall and less complex diffuser design. Additionally, the dense, negatively buoyant concentrate is countered by the warm, positively buoyant power plant discharge, further increasing the compatibility of the concentrate. This reduces the extent of the mixing zone, and reduces the amount of time needed for adequate mixing and dilution. The practice of co-location has been successfully implemented in Tampa Bay, Florida, which has an input of 166,000 m^3/day from the neighbouring Big Bend Power Station (Water Technology 2011).

Co-location is not be suitable for all desalination plants. This process only becomes feasible if the volume of cooling water discharged from the power plant is at least three to four times greater than the capacity of the desalination plant (Voutchkov 2004). Furthermore, corrosion from power plant heat exchangers may elevate the levels of metal in the feed to the desalination plant, which may then damage the reverse osmosis membrane units (Voutchkov 2004).

4.6 Regulations

Regulations vary according to the discharge location and the relevant governing body. The approval of a discharge permit will depend on a comparison between the water quality standard of the receiving body, and the quality of the concentrate (Mickley 2006). If regulatory requirements cannot be met upon discharge, a mixing zone may be defined. This is a quantified area or volume of water in which the water quality may exceed regulated standards. The monitoring of TDS and pollutant levels is required within the specified mixing zone. Discharge limits can be placed upon total suspended solids (TSS), TDS, salinity and other specific contaminants (Bergman 2007).

References

Alameddine, I., El-Fadel, M.: Brine discharge from desalination plants: a modeling approach to an optimized outfall design. Desalination 214(1–3), 241–260 (2007)

Al-Barwani, H., Purnama, A.: Re-assessing the impact of desalination plants brine discharges on eroding beaches. Desalination 204(1–3), 94–101 (2007)

Al-Barwani, H., Purnama, A.: Simulating brine plumes discharged into the seawaters. Desalination 221(1–3), 608–613 (2008)

Bergman, R.: Reverse Osmosis and Nanofiltration. American Water Works Association Denver (2007)

Bleninger, T., Jirka, G.: Modelling and environmentally sound management of brine discharges from desalination plants. Desalination 221(1−3), 585−597 (2008)

Cipollina, A., Bonfiglio, A., Micale, G., Brucato, A.: Dense jet modelling applied to the design of dense effluent diffusers. Desalination 167, 459−468 (2004)

Del Bene, J.V., Jirka, G., Largier, J.: Ocean Brine Disposal. Desalination 97(1−3), 365−372 (1994)

Höpner, T.: A procedure for environmental impact assessments (EIA) for seawater desalination plants. Desalination 124(1−3), 1−12 (1999)

Lattemann, S., Höpner, T.: Seawater Desalination: Impacts of Brine and Chemical Discharge on the Marine Environment. Desalination Publications, L'Aquila (2003)

Lattemann, S., Höpner, T.: Environmental impact and impact assessment of seawater desalination. Desalination 220(1−3), 1−15 (2008)

Malcangio, D., Petrillo, A.F.: Modeling of brine outfall at the planning stage of desalination plants. Desalination 254(1−3), 114−125 (2010)

Mauguin, G., Corsin, P.: Concentrate and other waste disposals from SWRO plants: characterization and reduction of their environmental impact. Desalination 182(1−3), 355−364 (2005)

Mickley, M.: Membrane Concentrate Disposal: Practices and Regulation, Desalination and Water Purification Research and Development Program Report No. 123 (Second Edition). U.S. Department of the Interior, Bureau of Reclamation, Denver (2006)

Mickley, M.: Treatment of Concentrate, Desalination and Water Purification Research and Development Program Report No. 155. U.S. Department of the Interior, Bureau of Reclamation, Denver (2009)

MixZon Inc: CORMIX Mixing Zone Model. http://www.mixzon.com/. Accessed 1 June 2011

Purnama, A., Al-Barwani, H.H.: Some criteria to minimize the impact of brine dischargeinto the sea. Desalination 171(2), 167−172 (2004)

Purnama, A., Al-Barwani, H.H., Al-Lawatia, M.: Modeling dispersion of brine waste discharges from a coastal desalination plant. Desalination 155, 41−47 (2003)

Schliephake, K., Brown, P., Mason-Jefferies, A., Lockey, K., Farmer, C.: Overview of Treatment Processes for the Production of Fit for Purpose Water: Desalination and Membrane Technologies, ASIRC Report No.: R05-2207. Australian Sustainable Industry Research Centre Ltd., Churchill (2005)

Sommariva, C.: Desalination and advanced water treatment : economics and financing. Balaban Desalination Publications, Hopkinton (2010)

United States Environmental Protection Agency: Visual Plumes. http://www.epa.gov/ceampubl/swater/vplume/. Accessed 1 June 2011

Voutchkov, N.: Seawater desalination costs cut through power plant co-location. Filtr. Sep. 41(7), 24−26 (2004)

Voutchkov, N., Sommariva, C., Pankratz, T., Tonner, J.: Desalination Process Technology. In: Cotruvo, J.A., Voutchkov, N., Fawell, J., Payment, P., Cunliffe, D., Lattemann, S. (eds.) Desalination Technology—Health and Environmental Impacts. CRC Press, Boca Raton (2010)

Water Technology: Tampa Bay Seawater Desalination Plant, Florida. http://www.water-technology.net/projects/tampa/. Accessed 28 June 2011

World Health Organisation: Desalination for Safe Water Supply, Guidance for the Health and Environmental Aspects Applicable to Desalination. World Health Organisation, Geneva (2007)

Watson, I.C., Morin, O.J., Jr., Henthorne, L.: Desalting Handbook for Planners, Third Edition, Desalination Research and Development Program Report No. 72. United States Department of the Interior, Bureau of Reclamation, Denver (2003)

Younos, T.: The Economics of Desalination. J. Contemp. Water Res. Edu. 132(1), 39−45 (2005)

Chapter 5
Deep Well Injection

Abstract Deep well injection is the disposal of concentrate into the voids and pores of rocks deep underground. Concentrate is injected down a well that consists of several layers of casing and grouting. Porous rocks are then used to contain the concentrate, while shale, clay and other impermeable rock formations are used to prevent the water contaminating aquifers. The conditions required for deep well injection are quite specific, and as such this disposal option is not widely employed.

Keywords Aquifer contamination · Confining layer · Costs · Environmental concerns · Receptor zones · Site selection · Well design · Well drilling

5.1 Design

An injection well is comprised of an injection tube, surrounded by layers of concentric casing with varying depths. A typical design for a deep injection well can seen in Fig. 5.1. The exact number of casings that are required and their depths will be determined by the geological conditions and surrounding aquifers.

To prevent fluid flowing back up the casing, the gap between the injection tubing and the innermost casing can be sealed with a packer (Shammas et al. 2009). Additionally, the gaps between the bore hole and the casing, and between layers of casing must be filled with cement to help maintain the structural integrity of the casing, to protect the casing from corrosion, and to prevent concentrate migrating into other aquifers (Shammas et al. 2009). Cementing is regarded as the most crucial factor in preventing contamination of potable aquifers, and thus testing of the cement to ensure a seal is of great importance (Mickley 2006). Upon completion of a well, a thorough inspection needs to be carried out to check the integrity of the well and to prevent failures. Depending on the positioning of aquifers and other subterranean

B. Ladewig and B. Asquith, *Desalination Concentrate Management*,
SpringerBriefs in Green Chemistry for Sustainability,
DOI: 10.1007/978-3-642-24852-8_5, © The Author(s) 2012

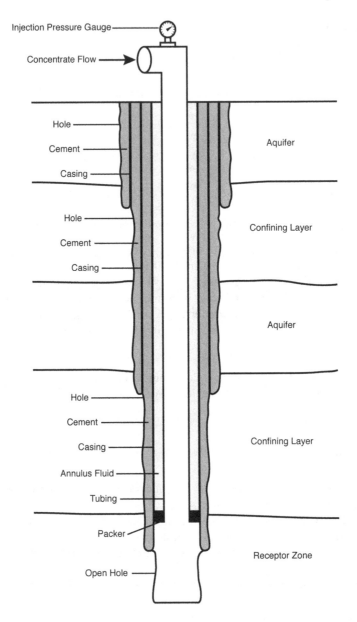

Injection Pressure Gauge

Concentrate Flow

Hole

Cement

Casing

Hole

Cement

Casing

Hole

Cement

Casing

Annulus Fluid

Tubing

Packer

Open Hole

Aquifer

Confining Layer

Aquifer

Confining Layer

Receptor Zone

Fig. 5.1 Typical injection well with tubing and three layers of casing

conditions, injection well depths typically range from 300 to 2400 m (Chelme-Ayala et al. 2009).

During operation, concentrate will undergo pretreatment before being pumped into the well. The injection pressure will vary with the depth of the well and the geological conditions of the well, and these can range from several atmospheres up to several hundred atmospheres (Mickley 2006). Further specific construction requirements will be needed based on the class of well to be constructed, as stipulated by the relevant regulatory body.

5.2 Site Selection

When selecting a site for an injection well, it is crucial to understand the hydro-geological conditions to ensure any existing or potential fresh water supplies do not become contaminated. Potable aquifers are more likely to be found at shallower depths, while deep aquifers are more likely to be lower in quality and are less likely to be used as a water source (Mickley 2006). Siting injection wells away from areas known to by seismically active also helps to reduce the possibility of contaminating aquifers. The use of injection wells should also be avoided near recoverable resources like ores, coal, oil and gas (Mickley 2006). The selected receiving aquifer must be able to contain the volume of concentrate for the expected life of the plant (Chelme-Ayala et al. 2009).

The economic and environmental suitability of any proposed well will be determined from surveying, testing, and the design of the well. When surveying and testing to determine site suitability, the confinement conditions, receptor zone and subsurface hydrodynamics should be studied with respect to the concentrate to be injected (Shammas et al. 2009).

5.2.1 Confinement Conditions

Confining layers are layers of shale, clay, or other rock formations with low permeability that slow the upwards flow of concentrate, rather than stopping it completely. An appropriate site for an injection well will have a confining layer with a suitably low leakage rate and fluid velocity. Shammas et al. (2009) give the following equations to determine the leakage rate and fluid velocity through a confining bed:

$$Q = PIA \tag{5.1}$$

$$v = \frac{PI}{\Phi} \tag{5.2}$$

where Q is the rate of leakage (m^3/day), v is the velocity of fluid (m/day), A is the leakage area (m^2), P is permeability (m^3/day.m^2), I is the hydraulic gradient (m/m) and Φ is the porosity of the confining layer.

Through sampling and testing of the confining layer with the concentrate that is to be injected, the rate of leakage, and hence the suitability of a confining layer, can be determined.

5.2.2 Receptor Zones

The receptor zone is the area into which the concentrate is injected. Any potential physical or chemical changes to the concentrate or rock strata in this zone due to chemical reactions with the concentrate need to be determined before injection begins. Failure to understand these changes may result in precipitation and plugging of pores, reducing the effectiveness of the well or even halting operation altogether (Shammas et al. 2009). Plugging in the injection zone can be prevented by understanding the compatibility of the rocks and fluids in the receptor zone with the concentrate. This includes understanding the pressure and temperature of the injection zone. During operation, an increase in pressure at the wellhead is an indicator that the aquifer cannot receive or transmit the concentrate.

5.2.3 Subsurface Hydrodynamics

Successful operation of an injection well relies on the prediction of the direction and rate of concentrate flow, the displacement of pre-existing water, and the aquifer pressure change over time can be estimated. A knowledge of the subsurface hydrodynamics is therefore required to assist in these predictions (Shammas et al. 2009).

5.3 Cost

Deep well injection has a strong economy of scale, and when geological conditions are favourable, it can be an economical method of concentrate management for larger desalination plants. Nevertheless, due to the large costs of drilling and installation, its use is limited to larger desalination plants.

5.3.1 Well Drilling and Formation

For other concentrate management options, the overall cost is largely determined by the flow rate of the concentrate. For the case of deep well injection however, the most costly factor is the drilling and formation of the well. This includes drilling, installation of tubing and casing, rig hire, surface equipment for waste disposal and continual surveying, logging and testing throughout construction. In particular it is the cost of labour rather than materials that increases the overall cost (Mickley 2006).

The drilling cost is determined by the depth of the well and the geology of the site. Tubing, casing and concreting costs all depend upon the depth and diameter of the well, and the selected material of construction (Mickley 2006). Since the flow rate of concentrate is less significant than the cost of construction, the diameter of the well can vary without significantly altering the cost. As a result of this, the well can be designed with larger than required diameters that allow for future increases in concentrate flow without greatly compromising the cost (Mickley 2006).

Various forms of logging, surveying and testing are undertaken throughout the drilling process. These are used to determine the profile of the rock formations surrounding the well, to determine the compatibility of the concentrate in the injection zone, and to check the integrity of the well, casing, cementing and tubing. A list of common logging, surveying and testing techniques is given in Table 5.1.

5.3.2 Surface Equipment

A pumping system is required to inject the concentrate into the well. The cost of pumps depends on the desired injection pressure and the volume of concentrate, and higher pressures and greater flow rates demand more costly pumps (Mickley 2006).

Pretreatment is required to remove suspended solids from the concentrate and to reduce the likelihood of precipitates forming in the injection zone. Solids may be present in the concentrate, especially when filter backwash is blended with the concentrate prior to injection. Pretreatment costs include filtration of the solids and chemicals for pH adjustment (Mickley 2006).

Throughout the life of the well there will be times when it must be taken offline for testing and maintenance. When this occurs, the plant must be able to facilitate the storage of concentrate, or have an alternative concentrate management option. These measures should be included in the total cost of the well.

5.3.3 Operation and Monitoring

Operating costs for injection wells are significantly lower when compared to capital costs. They include labour for operation and maintenance, chemicals for pretreatment, and power for pump operation (Mickley 2006). During operation of the well, the power required for pumping is the most significant cost (Mickley 2006).

Monitoring the well for leaks and nearby aquifers for contamination must also be considered. The extent of monitoring will vary based on the applicable environmental regulations and nearby potable aquifers. Monitoring wells can be constructed to monitor shallow or deep aquifers, and the cost of a these wells is determined by the required depth and the geology of the site. Additional costs include ancillary surface equipment and alarm systems (Mickley 2006).

Well monitoring systems are also used to detect failures in the injection well. This can be done in two ways; monitoring a confining layer between the injection zone and an aquifer, or monitoring of the annulus between the tubing and the innermost casing. To monitor the confining layer, the outer annulus is left exposed to the confining layer (i.e., no cementing), and a sample pipe extended from the surface to the aquifer can be used to monitor the fluid in the

Table 5.1 Sampling, logging, surveying and testing techniques performed during injection well drilling and construction

Technique	Description
Caliper surveying	Caliper testing measures the diameter of the borehole. These measurements can also be used to determine the mechanical integrity of the well and the volume of concrete that is required
Core sampling	Core samples from the well bore are taken to determine the rock formation and soil conditions. Samples can also be used to determine the flow potential of the receptor zone, concentrate compatibility, and the best drilling practices for the well
Gamma-ray log	This is a log of the natural formation radioactivity level. Gamma radiation is measured along the borehole to determine the rock and sediment formations
Geophysical logging	Records and profiles the rock formations surrounding the well
Hydrostatic pressure testing	The tubing annulus is filled with water to ensure its integrity and to check for leaks
Injection tests	Testing prior to concentrate injection to determine the flow rate and pressure at which the concentrate can be injected
Radioactive-tracer log	Tracer fluid movements are measured to produce a radioactive-tracer log. This shows the flow of fluid in the casing, tubing and the annulus, and helps to estimate flow rates, leaks, and other points of exit or entry for fluid into the borehole
Temperature log	This is a measurement of the temperature gradient throughout the well
Video surveying	Cameras are lowered down the well, which allows for a visual inspection of the well. In particular, the integrity of the borehole, casing, concrete and tubing can be checked. The fractures in the receptor zone into which the concentrate is injected can also assessed
Water sampling	Water samples from the injection zone are taken to predict the compatibility of the concentrate and the formation fluid

confining layer (Shammas et al. 2009). Any changes to the pressure or concentration of the fluid in the confining layer will indicate a possible leakage. Alternatively, the annulus between the tubing and the innermost casing is pressurised, or a low density fluid is inserted. A pressure change of this fluid can indicate a tubing leak or well failure before surrounding aquifers are contaminated. It must be noted that the pressure of this fluid can change due to other factors, including changes in the injection rate or temperature, and so these changes must be anticipated (Shammas et al. 2009).

5.4 Environmental Concerns

Due to the capacity of deep wells to store injected waste for a long period of time, if the correct measures are taken in design, construction and operation, deep well injection can provide an effective and environmentally safe method of concentrate management. The major environmental concern for deep well injection is the potential for contamination of nearby aquifers, which may be used as a source of drinking water. Six pathways have been defined that describe the potential migration of concentrate that can cause contamination of aquifers (Shammas et al. 2009; United States Environmental Protection Agency 2002):

(1) **Contamination via a failure in the injection well casing**

Failure of the injection well casing can cause concentrate to migrate from the well. This can be prevented through regular testing and maintenance of the casing, as well as appropriate well monitoring. In the event of casing failure, the use of tubing rather than casing for the innermost layer will reduce the severity by eliminating direct contact between the casing and concentrate. The use of tubing also helps to reduce casing corrosion.

(2) **Migration of concentrate upwards through the annulus between the casing and the borehole**

When resistance in the injection zone is too great, concentrate may flow upwards between the casing and the borehole rather than into the receptor. This can be prevented through adequate cementing of aquifers, and in particular cementing of the casing both directly above the injection zone, and directly below the lowest protected aquifer. The installation of a packer between the casing and the tubing will also help prevent the upwards migration of fluid.

(3) **Movement of concentrate through the confining zone**

Significant flow through the confining layer can occur if the confining layer has a greater permeability than expected, or if it contains fractures. Accurate predictions of the fluid movement are crucial in preventing significant flow through the confining zone, thus thorough studying and testing of the confining layer and injection zone should be undertaken. An understanding of the fluid flow rate at various pressures will prevent upwards flow caused by excessive injection pressure. Additionally, the use of deep receptor will provide greater protection for overlying aquifers by increasing the distance between the injection zone and other aquifers.

Fig. 5.2 Aquifer connected to the injection zone of a well with no barrier. Adapted from Shammas et al. (2009)

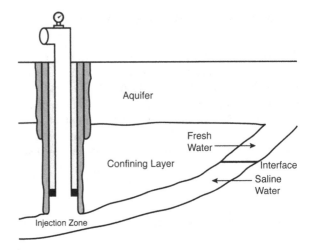

(4) **Movement of concentrate vertically through abandoned or completed wells which have not been correctly sealed**

When concentrate is injected into a receptor, an increase in pressure will be experienced in the surrounding area. Any existing or disused wells in this area need to be inspected to ensure they are adequately sealed and that their integrity is maintained. If this does not occur, there is potential for fluid to migrate into aquifers via these wells.

(5) **Lateral movement through the injection zone into areas that contain potable water sources**

It is possible for a potable water source to be connected to the injection zone with no confining layer or impermeable material as a barrier. An example of this can be seen in Fig. 5.2. Injection may continue if the concentrate flows away from the connected aquifer and there is no threat of contamination. Nonetheless, the prediction and knowledge of the movement and final location of the waste front is a crucial factor.

(6) **Direct injection into or above a potable aquifer**

When sufficient testing and surveying of the well site has not been undertaken, it is possible that the concentrate may be injected into or above an unknown potable aquifer. Thorough hydrogeological knowledge of area is therefore required to avoid such mistakes.

References

Chelme-Ayala, P., Smith, D.W., El-Din, M.G.: Membrane concentrate management options: a comprehensive critical review. Can. J. Civ. Eng. **36**(6), 1107–1119 (2009)

Mickley, M.: Membrane Concentrate Disposal: Practices and Regulation, Desalination and Water Purification Research and Development Program Report No. 123 (Second Edition). U.S. Department of the Interior, Bureau of Reclamation, Denver (2006)

Shammas, N.K., Sever, C.W., Wang, L.K.: Deep-Well Injection for Waste Management. In: Wang, L.K., Shammas, N.K., Hung, Y.-T. (eds.) Handbook of Environmental Engineering, Volume 9: Advanced Biological Treatment Processes. Humana Press, New York (2009)

United States Environmental Protection Agency: Technical Program Overview: Underground Injection Control Regulations. United States Environmental Protection Agency (2002)

Chapter 6
Spray Irrigation

Abstract Irrigated agriculture is the largest consumer of the world's water resources. A decline in the availability of fresh water is driving the use of waters with increasing salinity to be used for irrigation. The irrigation of salt tolerant plants, or halophytes, is a way of addressing this issue and freeing up fresh water for other uses. In the context of brine disposal, spray irrigation is the practice of using concentrate for irrigation of crops, lawns, and other vegetation. The major requirement of this process is that the salinity of the water is acceptable for use on the desired plant. Depending on the salinity of the concentrate, it may be used as is, or diluted with less saline water to bring the concentration of salt to a range where it is acceptable to use on halophytes and salt tolerant non-halophytes. As each plant species has a different salinity tolerance, the amount of concentrate that can be applied depends on the species of plant, the characteristics and salinity of the soil and concentrate that is to be used.

Keywords Conventional crop irrigation · Costs · Crop yield · Environmental concerns · Halophyte irrigation

6.1 Crop Irrigation

Dryland salinity is causing a decline in the productivity of land used for conventional agriculture. In spite of this, there is a range of halophytes and salt-tolerant non-halophytes that are capable of being used as crops in these salt-affected areas. Due to the lower salinity of concentrate from nanofiltration, microfiltration and ultrafiltration processes, spray irrigation is a more viable disposal option for these than for reverse osmosis processes. The successful use of membrane concentrate for crop irrigation requires the selection of appropriate vegetation. Either conventional crops or halophytes can be irrigated with membrane concentrate.

B. Ladewig and B. Asquith, *Desalination Concentrate Management,* 41
SpringerBriefs in Green Chemistry for Sustainability,
DOI: 10.1007/978-3-642-24852-8_6, © The Author(s) 2012

Areas best suited to irrigation with concentrate are those with level ground and warm climates. Nearby potable aquifers, unsuitable soil conditions and local regulations may altogether prevent spray irrigation at certain sites (Mickley 2009). Unfortunately the potential for spray irrigation as a method of managing concentrate is restricted, as large amounts of land are needed, and the geology must be suitable to prevent the accumulation of salt in the soil and the leaching of salt into underground sources of freshwater.

6.1.1 Halophyte Crop Irrigation

A growing number of studies have begun to focus on the use of concentrate from reverse osmosis plants for the irrigation of halophytes (Jordan et al. 2009; Riley et al. 1997; Soliz et al. 2011). Halophytes have possible uses as forages, oilseeds and biofuels (Glenn et al. 1991, 1998; Hendricks and Bushnell 2008; Rogers et al. 2005; Swingle et al. 1996) but more research is required to see these uses reach their full potential. Moreover, the characterisation of water consumption and growth of halophytes is required to be able to develop sustainable water management strategies (Jordan et al. 2009). Aronson (1989) has created a database listing 1560 known halophytes, but it is estimated that 5000–6000 halophyte species exist (Le Houerou 1993).

Halophytes may be grown in salt-affected areas, but to do this effectively, species need to be developed that are adaptable to various soil conditions, salinity and waterlogging stress levels, climates and livestock management systems (Rogers et al. 2005). For the successful use of halophytes as irrigated crops, the following conditions should be met (Glenn et al. 1999):

- Halophytes need to have a high yield potential
- The irrigation requirements of halophytes should be similar to that of conventional crops, and the irrigation must not damage the soil
- Halophyte products should be substitutable for conventional crop products
- High salinity agriculture needs to be able to be integrated with existing agricultural infrastructure

Research into the development of halophytes has largely focused on their use as animal fodder. Swingle et al. (1996) have demonstrated that halophytes can be incorporated into livestock feed without producing health problems or negatively affecting the carcass quality. When using halophytes as fodder, it is recommended that the use of halophyte forage accounts for no more than 50% of an animal's diet so that the loss of energy content in the salt-containing forage can be balanced by an increase in consumption (Swingle et al. 1996; Glenn et al. 1999). Substituting halophytes for a portion of animal fodder has also been found to decrease feed efficiency and increase animal water consumption. Although the fresh water requirements of animals may increase, this amount can easily be offset by the irrigation water saved when using desalination concentrate.

6.1.2 Conventional Crop Irrigation

The application of salt water for the irrigation of conventional crops has been the subject of a number of studies (Glenn et al. 1999; Grattan 2004; Grattan et al. 2004; Grieve et al. 2004; Skaggs et al. 2006a, b). However, there exists less potential for the irrigation of conventional crops with membrane concentrate when compared with halophytes. This is particularly a result of the limited range of conventional crop species that are tolerant to salt. Those which are most salt tolerant include sugarbeet, cotton, barley, date palms and some varieties of wheatgrass (Maas 1985). Saline irrigation waters and the presence of salt in the soil hinder the ability of most plants to absorb water, and most crops require irrigation with water containing a salt concentration of less than 7000 mg/L. Only the most tolerant are generally able to exceed this concentration (Rhoades et al. 1992). The irrigation of conventional crops will therefore usually require some degree of dilution, based on the salinity of the concentrate.

The yield of a crop irrigated with saline water is related to its salinity tolerance, such that once the threshold of tolerance is exceeded, the yield will decline. The following relationship derived by Mass and Hoffman (1977) shows this decrease in yield as soil salinity exceeds the threshold salinity for a species of plant:

$$Y_R = 100 - B(EC_e - A) \tag{6.1}$$

where Y_R is the relative yield of the crop, EC_e is the soil salinity (dS/m), A is the plant's salinity threshold (dS/m), the point after which the expected yield begins to decrease, and B is the decline in yield per unit decrease in salinity. Comprehensive data tables including the salinity threshold and decline in yield for a wide range of plant species can be found elsewhere (Maas 1985, 1990; Maas and Hoffman 1977). The values of salinity used for the calculation of the yield are based on water with high concentrations of sodium chloride, as opposed to other ions that may be present in concentrate. The calculated relative yield should be considered as an estimate only, as factors such as climate and soil conditions affect the threshold limit and decline in yield. A number of plant species are also less tolerant to salinity during germination, or throughout the emergence and seedling stages. Note that the amount of salt that is present in the root zone may alter the time a crop takes to reach maturity (Shannon et al. 1994).

Figure 6.1 is a plot of relative crop yield versus soil salinity, and shows the regions for classifying crops based on their salinity tolerance. The crop yield for barley is indicated by the dashed line, with its crop yield indicating that it is tolerant to soil salinity.

When irrigating crops with saline water, a number of management techniques can be employed to help offset the potential loss of yield. Meiri and Plaut (1985) describe a number of strategies to manage crops under saline conditions, including:

Fig. 6.1 Classifications of crop tolerance based on relative crop yield versus soil salinity. The tolerance of barley is shown by the dashed line. Adapted from Maas (1985) and Maas and Hoffman (1977)

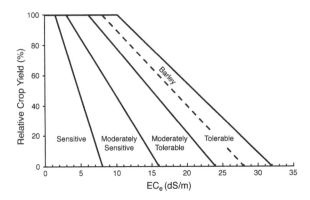

Controlling Root Zone Salinity

The salinity of the root zone can be controlled through numerous means, including adjusting the leaching fraction, adjusting the soil surface contour and seedling planting position, optimising the irrigation interval, using water of varying salinity at different growth stages, and blending water with different salinities. Waters with different salinities may be blended prior to irrigation or applied directly to crops and allowed to mix in the soil.

Optimising of Row Spacing

The field yield of a crop is determined by the yield of each plant, as well as each plant's density or stand (plants per unit area). Optimisation of the row spacing can increase the stand without negatively affecting the plant yield, hence increasing the field yield (Keren et al. 1983).

Reducing the Salinity Damage of the Plant

Reducing the damage caused by salinity will result in an improved crop yield. This can be done through changes in crop management or environmental conditions. Specifically, changes in irrigation method (Maas 1985), climate (Hoffman and Jobes 1978; Hoffman and Rawlins 1971; Magistad et al. 1943) and ambient CO_2 concentration (Schwarz and Gale 1984) have been shown to affect the growth of crops in saline conditions.

6.2 Costs

Typical costs for spray irrigation include the pretreatment of concentrate, equipment required for the storage, blending and distribution of water, land preparation (including drainage systems) and the cost of land. Large tracts of lands are usually required to dispose of the concentrate, especially considering any increase in concentrate volume due to dilution (Mickley 2009). Spray irrigation is therefore best suited to smaller desalination plants with low volumes of concentrate. Unless additional concentrate management schemes are implemented, plants with concentrate volumes of greater than 0.4 million gallons per day (MGD) are usually considered unsuitable (Mickley 2009). High leaching fractions can also lead to greater water requirements and greater pumping costs.

6.3 Environmental Concerns

Anthropogenic salinity is largely due to poor irrigation management. As the use of concentrate for spray irrigation allows for the potential build up of salt in the soil and contamination of groundwater, adequate irrigation management is essential to prevent further increases in salinity.

Increases in soil salinity can hinder the growth of plants by limiting their uptake of nutrients (Grattan and Grieve 1992). The major cations that affect saline soils are Na^+, Ca^{2+}, Mg^{2+}, as well K^+, while the major anions are Cl^-, SO_4^{2-}, HCO_3^- and NO_3^-. When the soil has a high pH, CO_3^{2-} is also present. The nutrients which plants require most are Ca^{2+}, Mg^{2+} and K^+. However, the uptake of K^+ and Ca^{2+} is hindered by the presence of excess Na^+, and an increased concentration of Ca^{2+} in the soil can lead to a deficiency of Mg^{2+} (Grattan and Grieve 1992).

Accumulation of salt in the root zone can be prevented by allowing a certain amount of water to percolate through the root zone. This is known as the leaching requirement, and is defined as the leaching fraction, LF, such that:

$$LF = \frac{\text{amount of water that passes through the root zone}}{\text{amount of irrigation water applied}} \qquad (6.2)$$

This can also be expressed in terms of the surface depth of water, or the volume of water per unit surface area. The excess water that passes through is used to carry excess salts that can hinder growth. From the above equation, it can be seen that the greater the leaching fraction, the lower the salinity in the root zone but the greater the irrigation water requirement. Therefore, as the salinity of the concentrate increases, more water must be applied to remove the extra salt. The amount of salt that leaves the root zone may be greater than, equal to or less than the amount of salt in the irrigation water applied, based on precipitation and dissolution reactions of salts (Oster and Rhoades 1990). Several steady state and transient models exist for the estimation of leaching requirement. These can be found elsewhere (Corwin et al. 1990, 2007; Letey et al. 1985; Rhoades 1974; Rhoades and Merrill 1976; Richards 1954; Šimůnek and Suarez 1994).

Irrigation drainage from saline water must not leach into underlying potable aquifers. To do this, the site that is selected for irrigation should have a deep soil profile and should not be in a flood plain. These measures, along with irrigation management to minimise deep leaching and preferential flow, can help to prevent the contamination of potable aquifers (Riley et al. 1997).

References

Aronson, J.A.: HALOPH: A Data Base of Salt Tolerant Plants of the World. Office of Arid Land Studies, University of Arizona, Tucson (1989)

Corwin, D.L., Waggoner, B.L., Rhoades, J.D.: A functional model of solute transport that accounts for bypass. J. Environ. Qual. **20**, 647–658 (1990)

Corwin, D., Rhoades, J., Šimůnek, J.: Leaching requirement for soil salinity control: Steady-state versus transient models. Agric. Water Manag. **90**(3), 165–180 (2007)

Glenn, E.P., O'Leary, J.W., Watson, M.C., Thompson, T.L., Kuehl, R.O.: Salicornia bigelovii Torr.: An Oilseed Halophyte for Seawater Irrigation. Science **251**(4997), 1065–1067 (1991)

Glenn, E., Tanner, R., Miyamoto, S., Fitzsimmons, K., Boyer, J.: Water use, productivity and forage quality of the halophyte Atriplex nummularia grown on saline waste water in a desert environment. J. Arid Environ. **38**(1), 45–62 (1998)

Glenn, E.P., Brown, J.J., Blumwald, E.: Salt Tolerance and Crop Potential of Halophytes. Crit. Rev. Plant Sci. **18**(2), 227–255 (1999)

Grattan, S.: Evaluation of salt-tolerant forages for sequential water reuse systems: I. Biomass production. Agric. Water Manag. **70**(2), 109–120 (2004)

Grattan, S.R., Grieve, C.M.: Mineral element acquisition and growth response of plants grown in saline environments. Agric. Ecosyst. Environ. **38**(4), 275–300 (1992)

Grattan, S., Grieve, C., Poss, J., Robinson, P., Suarez, D., Benes, S.: Evaluation of salt-tolerant forages for sequential water reuse systems: III. Potential implications for ruminant mineral nutrition. Agric. Water Manag. **70**(2), 137–150 (2004)

Grieve, C., Poss, J., Grattan, S., Suarez, D., Benes, S., Robinson, P.: Evaluation of salt-tolerant forages for sequential water reuse systems II. Plant–ion relations. Agric. Water Manag. **70**(2), 121–135 (2004)

Hendricks, R.C., Bushnell, D.M.: Halophytes Energy Feedstocks: Back To Our Roots. In: The 12th International Symposium on Transport Phenomena and Dynamics of Rotating Machinery, Honolulu, Hawaii 2008

Hoffman, G.J., Jobes, J.A.: Growth and Water Relations of Cereal Crops as Influenced by Salinity and Relative Humidity1. Agron. J. **70**(5), 765–769 (1978)

Hoffman, G.J., Rawlins, S.L.: Growth and Water Potential of Root Crops as Influenced by Salinity and Relative Humidity1. Agron. J. **63**(6), 877–880 (1971)

Jordan, F.L., Yoklic, M., Morino, K., Brown, P., Seaman, R., Glenn, E.P.: Consumptive water use and stomatal conductance of Atriplex lentiformis irrigated with industrial brine in a desert irrigation district. Agric. For. Meteorol. **149**(5), 899–912 (2009)

Keren, R., Meiri, A., Kalo, Y.: Plant spacing effect on yield of cotton irrigated with saline waters. Plant and Soil **74**(3), 461–465 (1983)

Le Houerou, H.N.: Salt-tolerant plants for the arid regions of the Mediterranean isoclimatic zone. In: Lieth, H., Masoom, A. (eds.) Towards the Rational Use of High Salinity Tolerant Plants, vol. 1. pp. 403–422. Kluwer Academic Publishers, Dordrecht (1993)

Letey, J., Dinar, A., Knapp, K.C.: Crop-Water Production Function Model for Saline Irrigation Waters. Soil Sci. Soc. Am. J. **49**(4), 1005–1009 (1985)

Maas, E.V.: Crop tolerance to saline sprinkling water. Plant and Soil **89**(1), 273–284 (1985)

Maas, E.V.: Crop salt tolerance. In: Tanji, K.K. (ed.) Agricultural Salinity Assessment and Management. Manuals and Reports on Engineering No. 71, pp. 262–304. American Society of Civil Engineers, New York (1990)

Maas, E.V., Hoffman, G.J.: Crop salt tolerance—current assessment. J. Irrig. Drain. Div. **103**(2), 115–134 (1977)

Magistad, O.C., Ayers, A.D., Wadleigh, C.H., Gauch, H.G.: Effect of salt concentration, kind of salt, and climate on plant growth in sand cultures. Plant Physiol. **18**(2), 151–166 (1943)

Meiri, A., Plaut, Z.: Crop production and management under saline conditions. Plant and Soil **89**(1), 253–271 (1985)

Mickley, M.: Treatment of Concentrate, Desalination and Water Purification Research and Development Program Report No. 155. U.S. Department of the Interior, Bureau of Reclamation, Denver (2009)

Oster, J.D., Rhoades, J.D.: Steady-State Root Zone Salt Balance. In: Tanji, K.K. (ed.) Agricultural Salinity Assessment and Management. American Society of Civil Engineers, New York (1990)

Rhoades, J.D.: Drainage for Salinity Control. In: van Schilfgaarde, J. (ed.) Drainage for Agriculture. Agronomy Monograph No. 17. SSSA, Madison (1974)

Rhoades, J.D., Merrill, S.D.: Assessing the suitability of water for irrigation: theoretical and empirical approaches. In: Prognosis of Salinity and Alkalinity. Food and Agriculture Organization of the United Nations, Rome (1976)

Rhoades, J.D., Kandiah, A., Mashali, A.M.: The use of saline waters for crop production (FAO irrigation and drainage paper 48). Food and Agriculture Organization of the United Nations, Rome (1992)

Richards, L.A. (ed.) Diagnosis and Improvement of Saline and Alkali Soils. United States Department of Agriculture, Washington, D.C. (1954)

Riley, J.J., Fitzsimmons, K.M., Glenn, E.P.: Halophyte irrigation: an overlooked strategy for management of membrane filtration concentrate. Desalination **110**, 197–211 (1997)

Rogers, M.E., Craig, A.D., Munns, R.E., Colmer, T.D., Nichols, P.G.H., Malcolm, C.V., Barrett-Lennard, E.G., Brown, A.J., Semple, W.S., Evans, P.M., Cowley, K., Hughes, S.J., Snowball, R., Bennett, S.J., Sweeney, G.C., Dear, B.S., Ewing, M.A.: The potential for developing fodder plants for the salt-affected areas of southern and eastern Australia: an overview. Aust. J. Exp. Agric. **45**(4), 301–329 (2005)

Schwarz, M., Gale, J.: Growth Response to Salinity at High Levels of Carbon Dioxide. J. Exp. Bot. **35**(2), 193–196 (1984)

Shannon, M.C., Grieve, C.M., Francois, L.E.: Whole-plant response to salinity. In: Wilkinson, R.E. (ed.) Plant-Environment Interactions. Marcel Dekker, New York (1994)

Šimůnek, J., Suarez, D.L.: Two-dimensional transport model for variably saturated porous media with major ion chemistry. Water Resour. Res. **30**(4), 1115–1133 (1994)

Skaggs, T.H., Poss, J.A., Shouse, P.J., Grieve, C.M.: Irrigating Forage Crops with Saline Waters: 1. Volumetric Lysimeter Studies. Vadose Zone J. **5**(3), 815–823 (2006a)

Skaggs, T.H., Shouse, P.J., Poss, J.A.: Irrigating Forage Crops with Saline Waters: 2. Modeling Root Uptake and Drainage. Vadose Zone J. **5**(3), 824–837 (2006b)

Soliz, D., Glenn, E.P., Seaman, R., Yoklic, M., Nelson, S.G., Brown, P.: Water consumption, irrigation efficiency and nutritional value of *Atriplex lentiformis* grown on reverse osmosis brine in a desert irrigation district. Agric. Ecosyst. Environ. **140**(3–4), 473–483 (2011)

Swingle, R.S., Glenn, E.P., Squires, V.: Growth performance of lambs fed mixed diets containing halophyte ingredients. Anim. Feed Sci. Technol. **63**(1–4), 137–148 (1996)

Chapter 7
Evaporation Ponds

Abstract Evaporation ponds are a low technology solution for concentration management whereby brine is pumped into large ponds and water slowly evaporates via direct solar energy. They are a widely used method of saline water management in the Middle East and Australia. The simplicity of the process reduces maintenance and operating costs, and greatly reduces energy requirements. Leakage from evaporation ponds can cause contamination of surrounding land and water sources, so the prevention of leakage through adequate design is crucial. Evaporation ponds should not be confused with solar gradient ponds, which use concentrated brine to produce electricity.

Keywords Costs · Environmental concerns · Evaporation pond design · Evaporation rate · Liners · Pond area · Pond banks · Pond depth · Social impacts

7.1 Design

The design of an evaporation pond must take into account both the volume of concentrate from the plant and the evaporation rate at the selected site. The prevention of salinity in surrounding areas and the contamination of nearby potable aquifers is of great importance and must be carefully considered during the design phase. The major design factors to consider are the pond area, depth, liners and bank size.

7.1.1 Pond Area

The required surface area of a pond is dependent upon both the volume of concentrate and the evaporation rate. Surface area, A (m^2), can be estimated using the following equation:

B. Ladewig and B. Asquith, *Desalination Concentrate Management*, 49
SpringerBriefs in Green Chemistry for Sustainability,
DOI: 10.1007/978-3-642-24852-8_7, © The Author(s) 2012

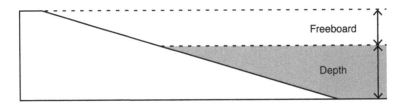

Fig. 7.1 Cross sectional view of the inner bank of an evaporation pond showing depth and freeboard

$$A = \frac{V}{E} \tag{7.1}$$

where V is the volume of concentrate sent to the pond (m³/day) and E is the evaporation rate (m/day). A greater volume of concentrate will increase the required pond area, and a higher evaporation rate will reduce the required area. Laser leveling can be used to ensure a more even evaporation rate across the pond, and hence a smaller required area (Ahmed et al. 2000). A safety factor should also be applied to the calculated pond size for the event of a lower than anticipated evaporation rate, or unexpected increases in concentrate volume. It is recommended that a 20% safety factor be applied to surface area of the pond (Mickley 2006).

7.1.2 Pond Depth

The minimum depth of a pond is directly proportionate to the rate of evaporation. This depth needs to allow for increases in volume, the precipitation of salts, as well as for rainfall and waves. It is estimated that the best evaporation rate can be achieved with pond depths of 0.03–0.45 m, however ponds with depths of up to 1.02 m have been shown to have reasonable evaporation rates (Mickley 2006). Similarly to pond area, a safety factor can be applied to the calculated minimum pond depth to increase the capacity and prevent the pond from overflowing. This extra depth will depend upon the expected additional discharge volume at the beginning of plant operation (Mickley 2006), and the ambient conditions during winter, at which time the pond may store water rather than reduce its volume (Ahmed et al. 2000).

The freeboard of an evaporation pond is the height between the design depth and the top of the bank, as shown in Fig. 7.1. Freeboard accounts for wave action due to wind, volume increases when the evaporation rate is lower than expected, surges and rainfall. Ahmed et al. recommend that the freeboard height be 0.2 m (Ahmed et al. 2000), however recommendations for evaporation basins on the Riverine Plain in Australia suggest freeboard should be 0.6–0.8 m (Leaney and Christen 2000). Ultimately, the freeboard height should be determined based on the specific climatic conditions at the site of the pond, including the average rainfall and anticipated wave height. The following equation may be used to estimate the height of pond wave action (Mickley 2006):

$$H = 0.007W\sqrt{F} \qquad (7.2)$$

where H is the wave height (m), W is the wind velocity (km/h) and F is the fetch, the straight line distance wind can blow without obstruction (km). Note that wave height can be minimised by constructing the pond with its length perpendicular to the prevailing wind direction (Ahmed et al. 2000).

7.1.3 Evaporation Rate

While the volume of concentrate can be determined based on plant capacity and recovery, the evaporation rate at any given site varies with climate. To determine the evaporation rate of fresh water at certain locations, a standard pan evaporation measurement is taken. Evaporation pans are small, open air pans filled with water from which losses in water due to evaporation are measured. Standard size Class A evaporation pans are most commonly used, which are 1.207 m in diameter and 0.25 m in depth. The daily change in depth, minus any rainfall, is used to determine the evaporation rate in mm/day. This rate takes into account the effects of climate on evaporation rate, but corrections for pond area and salinity must be made when determining the evaporation rate of a specific evaporation pond.

The humidity gradient between the water in the pond and the water in the air above the pond is the driving force for the evaporation process. Larger ponds have a different environment and greater humidity above the water, reducing this driving force and hence the evaporation rate (Leaney and Christen 2000; Morton 1986). The evaporation factor, F_1, is a correction factor that adjusts the pan evaporation rate based on pond size. It can be approximated by (Jolly et al. 2000):

$$F_1 = 1 - 0.029 \ln(A) \qquad (7.3)$$

where A is the pond area (ha). This relationship can be seen in Fig. 7.2. As the pond size approaches the size of the evaporation pan, the evaporation rate approaches that of the pan.

Evaporation rate is also affected by the salinity of the water. Saline water has a lower evaporation rate than fresh water, as salt reduces the vapour pressure of the water, decreasing the driving force for evaporation. A second evaporation factor, F_2, the evaporation rate of saline water compared to fresh water, can be approximated as (Leaney and Christen 2000):

$$F_2 = 1.025 - 0.0245e^{0.008795} \qquad (7.4)$$

where S is the salinity of the solution (g/L) up to 320 g/L. This relationship is shown in Fig. 7.3. It can be shown from this equation that the evaporation rate will

Fig. 7.2 Decreasing
evaporation rate due to the
oasis effect as pond size
increases. Adapted from Jolly
et al. (2000)

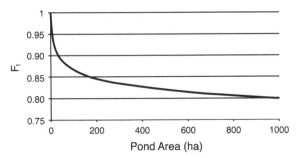

Fig. 7.3 Decreasing
evaporation rate of salt water
compared with fresh water as
salinity increases. Adapted
from (Leaney and Christen
2000)

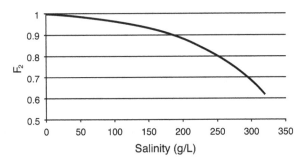

slow with time as water is evaporated and the salinity of the pond increases. Note
that when determining the evaporation rate for a pond, this relationship should be
used as an estimation only, as the exact relationship between salinity and evaporation rate relies on a great number of site specific factors other than salinity,
including air temperature, wind velocity, relative humidity, barometric pressure,
water surface temperature, heat exchange rate with the atmosphere, incident solar
absorption and reflection, pond thermal currents and pond depth (Mickley 2006).

When determining the evaporation rate for an evaporation pond in the absence
of a corrected pan evaporation rate, an approximation of 70% of the freshwater
evaporation rate is considered reasonable (Mickley 2006).

7.1.4 Pond Liners

Pond liners are used to prevent the leakage of concentrate, which can cause soil and
groundwater contamination. They are geomembranes fabricated from materials such
as high-density polyethylene, low-density polyethylene, polyvinyl chloride or
polypropylene. They should be strong enough to withstand cleaning without causing
leaks, and the pond needs to be deep enough to prevent the liner from drying and
cracking (Mickley 2006). The requirement for liners is a major drawback for
evaporation ponds, as they are usually the largest cost (Nicot et al. 2009).

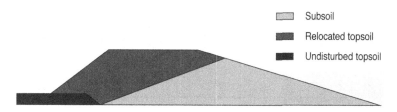

Fig. 7.4 Cross-sectional view of a bank showing a soil pattern to minimise erosion and encourage vegetation growth on the outer slope. Adapted from (Singh and Christen 2000)

The sealing of liner joints upon installation is very important, as this is where leaks are likely to occur (Glater and Cohen 2003). Several research efforts have investigated the possibility of 'self-sealing' evaporation ponds for concentrate disposal in Texas (Nicot et al. 2009; Turner et al. 1999). Self-sealing liners are created by adding chemicals to the concentrate that reduce the porosity and permeability of the existing liner or soil. Nicot et al. (2009) produced a report outlining possible approaches for self-sealing liners. Self-sealing liners have the potential to reduce costs by waiving regulations that stipulate the required thickness of liners and the groundwater monitoring systems. A reduction in liner thickness or ponds with no liners would significantly reduce the capital cost of a pond. Unfortunately, no practical self-sealing solution currently exists, and further work is required to reduce costs and produce a solution that is technically and economically viable.

Evaporation ponds or disposal basins are widely used in the Murray-Darling Basin in Australia to reduce salinity in irrigation areas. These basins are similar to evaporation ponds for concentrate disposal, but they may be designed to have a controlled amount of leakage. This eliminates the need for liners, but additional costly earthwork must be done to compact the soil and reduce its permeability, and additional leakage interception processes must be installed. Nonetheless, if a controlled amount of leakage is acceptable, this option is often cheaper than the use of a liner (Singh and Christen 2000). Specific guidelines for the construction and use of evaporation ponds for irrigation water can be found elsewhere (Jolly et al. 2000; Singh and Christen 2000; Christen et al. 1999).

7.1.5 Pond Banks

The banks around evaporation ponds can be built from the existing soil and excavated earth. It is suggested that a layer the topsoil be removed from the pond area, and the subsoil underneath be used to form the inside of the bank. The outer slope of the bank can be covered in the removed topsoil, as this promotes the regrowth of vegetation (Singh and Christen 2000). An example of this is shown in Fig. 7.4.

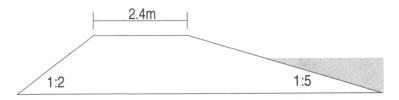

Fig. 7.5 Cross-sectional view of a bank showing suggested dimensions. Adapted from (Singh and Christen 2000)

Suggested dimensions for evaporation ponds are given by Singh and Christen (2000), and these can be seen in Fig. 7.5. The banks surrounding the pond should be a minimum of 1 m high and 2.4 m wide to allow for vehicular access. To reduce erosion, it is recommended that the slope of the inner bank should be 1:5 and the outside bank a slope of 1:2.

7.2 Pond Costs

The feasibility of an evaporation pond is determined by the volume of concentrate, the cost of land and the ambient evaporation rate. They are most cost-effective in areas with low rainfall, high evaporation rates, and where large expanses of land are available at low cost (Mickley 2009). This best suits inland desalination plants in regional and remote areas where these conditions can be met.

Evaporation ponds have a poor economy of scale and while they are economic for small waste flows, the largest feasible volume of concentrate is typically no greater than 5 MGD (Glater and Cohen 2003). Moreover, if the evaporation rate is low during the cooler months, the pond area may increase to an unfeasible size. In such instances, alternative disposal methods or concentrate storage options should be considered (Mickley 2009).

The major cost factors of an evaporation pond include pond liners, land preparation, excavation and clearing, site surveying, bank construction, pumps, control systems, disposal of precipitated solids, maintenance and geotechnical investigation of the site (Singh and Christen 2000). Of these, pond liners typically represent the greatest cost (Nicot et al. 2009).

Typical maintenance costs for an evaporation pond are pump maintenance and replacement, erosion control (including the replacement of eroded banks), vegetation management, wildlife control and seepage control. Bank rebuilding can include up to 10% of the banks every 10 years (Singh and Christen 2000). Annual operating costs are generally very low, and can be approximated at 0.5% of the total installation cost (Mickley 2006). The cost of the bank surrounding the pond increases with height, but so too does the storage capacity per unit area. A suitable balance must be found that allows enough freeboard and additional storage capacity while keeping construction costs to a minimum. Approximate bank, liner, fence and road costs based on pond size can be found elsewhere (Mickley 2006).

The disposal of solids is generally not required but may be necessary if the amount of precipitated solids becomes too great, if there is a high amount of suspended solids or dirt, or if the pond is too shallow (Mickley 2006). Salt disposal includes disposal to sea, landfill or other designated waste disposal sites, and this can become a costly process.

7.3 Environmental Concerns

The leakage of concentrate and potential contamination of the surrounding environment and aquifers with salt and other chemicals is the largest threat posed to the environment by evaporation ponds. There is also potential for the high salinity water to impact local wildlife and for salt spray to harm surrounding flora (Mickley 2009; Schliephake et al. 2005).

The contamination of groundwater can negate any existing use that water may have, or prevent its future use. Furthermore, if not prevented and controlled, salinisation can affect nearby crops, pastures, infrastructure and buildings. The impacts of salinity have been widely studied and will not be discussed here.

A number of measures can but put in place to monitor for leaks and help prevent the contamination of groundwater. These can also reduce the potential buildup in soil and groundwater of heavy metals which may be present in the concentrate, and which can lead to long-term problems (Mohamed et al. 2005). Monitoring wells can be installed to observe the water quality in nearby aquifers, and any changes in these may indicate a leak. When ponds have two liners (as may be required based on the extent of nearby aquifers and local regulations), any leakage that makes its way through the first liner may be detected via a connected monitoring sump, or through a moisture detector inserted between the two liners (Mickley 2006).

In some areas it may be acceptable for a small amount of leakage to occur from a pond, however this is more applicable for evaporation ponds used for irrigation water control rather than concentrate disposal. In such cases, the salt not contained by the pond should be contained in the soil and aquifer system directly below and adjacent to the pond (Christen et al. 1999). This ensures a large amount of the salt can eventually be recycled back into the evaporation ponds through appropriate interception methods. As stated above, leakage should not contaminate any groundwater with existing or potential use, and should minimise any potential environmental damage. The decision to allow leakage needs to be done on a case by case basis after an appropriate geological assessment.

To reduce the environmental impact of an evaporation pond, the environmental sensitivity, hydrogeology and land characteristics of the site should be assessed during site selection. In particular, the proximity of the site to conservational areas, flood plains, wetlands, swamps and residential and commercial areas needs to be assessed (Singh and Christen 2000). Once a site has been selected, detailed on-site hydrogeology assessment can be performed when required, including the depth, extent, piezometric level, transmissivity and water quality of any nearby aquifers (Singh and Christen 2000).

7.4 Social Impacts

Evaporation ponds are often viewed negatively, particularly due to potential salinisation of local land, unpleasant odours and aesthetic problems (Christen et al. 1999). The siting and design of an evaporation pond must then take these factors into account. The planting of trees around the perimeter of a disposal basin has been suggested as a way to increase social acceptance. The pond site should also include a buffer zone to position the pond an appropriate distance away from residential and commercial areas, schools, hospitals and other public areas (Jolly et al. 2000; Christen et al. 1999).

References

Ahmed, M., Shayya, W.H., Hoey, D., Mahendran, A., Morris, R., Al-Handaly, J.: Use of evaporation ponds for brine disposal in desalination plants. Desalination **130**(2), 155–168 (2000)

Christen, E., Jolly, I., Leaney, F., Narayan, K., Walker, G.: On-farm and community-scale salt disposal basins on the Riverine Plain : underlying principles for basin use. CRC for Catchment Hydrology, Clayton (1999)

Glater, J., Cohen, Y.: Brine Disposal From Land Based Membrane Desalination Plants: A Critical Assessment. University of California, Los Angeles (2003)

Jolly, I., Christen, E., Gilfedder, M., Leaney, F., Trewhella, B., Walker, G.: On-farm and community-scale salt disposal basins on the Riverine Plain: Guidelines for basin use. CRC for Catchment Hydrology, Clayton (2000)

Leaney, F., Christen, E.: On-farm and community-scale salt disposal basins on the Riverine Plain: evaluating basin leakage rate, disposal capacity and plume development. CRC for Catchment Hydrology, (2000)

Mickley, M.: Membrane Concentrate Disposal: Practices and Regulation, Desalination and Water Purification Research and Development Program Report No. 123 (Second Edition). U.S. Department of the Interior, Bureau of Reclamation, Denver (2006)

Mickley, M.: Treatment of Concentrate, Desalination and Water Purification Research and Development Program Report No. 155. U.S. Department of the Interior, Bureau of Reclamation, Denver (2009)

Mohamed, A.M.O., Maraqa, M., Al Handhaly, J.: Impact of land disposal of reject brine from desalination plants on soil and groundwater. Desalination **182**(1–3), 411–433 (2005)

Morton, F.I.: Practical Estimates of Lake Evaporation. J. Clim. Appl. Meteorol. **25**(3), 371–387 (1986)

Nicot, J.-P., Gross, B., Walden, S., Baier, R.: Self-Sealing Evaporation Ponds for Desalination Facilities in Texas. Texas Water Development Board, Austin (2009)

Schliephake, K., Brown, P., Mason-Jefferies, A., Lockey, K., Farmer, C.: Overview of Treatment Processes for the Production of Fit for Purpose Water: Desalination and Membrane Technologies, ASIRC Report No.: R05-2207. Australian Sustainable Industry Research Centre Ltd., Churchill (2005)

Singh, J., Christen, E.: On-Farm and Community-Scale Salt Disposal Basins on the Riverine Plain: Minimising the cost of basins: siting design and construction factors. CRC for Catchment Hydrology, (2000)

Turner, C.D., Walton, J.C., Moncada, J.D., Tavares, M.: Brackish Groundwater Treatment and Concentrate Disposal for the Homestead Colonia, El Paso, Texas. U.S. Department of the Interior, Bureau of Reclamation, Denver (1999)